The
PERMACULTURE
PROMISE

The PERMACULTURE PROMISE

What Permaculture Is and How It Can Help Us

Reverse Climate Change

Build a More Resilient Future on Earth

Revitalize Our Communities

JONO NEIGER

Foreword by Toby Hemenway

Storey Publishing

The mission of Storey Publishing is to serve our customers by publishing practical information that encourages personal independence in harmony with the environment.

Edited by Carleen Madigan
Art direction and book design by Jeff Stiefel
Indexed by Catherine Goddard

Front cover photograph © alvinge/123RF
Interior photography by © Alice Maggio, for BerkShares, Inc., 120; © Amanda Tipton Photography/Moment/Getty Images, 44; © Amos Chapple/Lonely Planet Images/Getty Images, 134; © Bernhard Lang/Stone/Getty Images, 68; © Bill Barksdale/Design Pics/Getty Images, 16–17; © blickwinkel/Alamy, 79; © Bloomberg/Getty Images, 56; © Brian Jolley, 9; © Bryan Mullennix/Tetra Images/Getty Images, 15; © Colin Monteath/Hedgehog House/Minden Pictures/Getty Images, 10 top; © Corbis Super RF/Alamy, 38; © Cultura Creative (RF)/Alamy, 76; courtesy of David Bainbridge, 111 bottom left; © David Hiser/National Geographic Magazines/Getty Images, 49; © David Speight/Alamy, 85; © Debra Behr/Alamy, 99; © E.J. Peiker/BIA/Minden Pictures/Getty Images, 51 bottom; © Environmental Education Media Project, 25; © Ethan Thompson, 36–37; FEMA/Joselyne Augustino/Wikimedia Commons, 145; © Gary Crabbe/Enlightened Images/Alamy, 46–47, 51 top; © Greg Wright/Alamy, 133 right; © Hugh Clark/FLPA/Minden Pictures, 21; © Images from BarbAnna/Moment/Getty Images, 64; © Jim West/Alamy, 43; © John Rensten/DigitalVision/Getty Images, 122; © Jono Neiger, 32, 57, 59, 60, 61, 63, 65, 66, 67, 70 top and bottom left, 71, 80, 86, 88, 90, 97, 101, 103, 104, 105, 111 bottom right, 112 left, 113 top and bottom right, 114 top, 136, 138, 144, 147, 148; © Kathy West Studios, 28, 29; © Kunal Mehra/Moment Open/Getty Images, 132–133; © Don Kelsen/Los Angeles Times/Getty Images, 111 top right; © Leonie Lambert/Photolibrary/Getty Images, 131; photo by Lynn Betts, USDA Natural Resources Conservation Service, 129; © magdasmith/E+/Getty Images, 62; © Mark Steinmetz/VISUM/The Image Works, 72; Mars Vilaubi, 2, 5, 6, 23–24, 48 top, 83 left, 92, 95, 112 right, 113 bottom left, 114 bottom, 117, 126, 154; © Michael Wheatley/Alamy, 27; © National Geographic Creative/Alamy, 26; © Otto Hahn/Picture Press/Getty Images, 41; © Panoramic Images/Getty Images, 50; © Paul Gordon/Alamy, 52; © Plantography/Alamy, 34; © Portland Press Herald/Getty Images, 89; © Richard Levine/Alamy, 55; © Sean Malyon/Photolibrary /Getty Images, 106; courtesy of Sim Van der Ryn, 58; © SONER BAKIR/Alamy, 40; © Stephen Shepherd/Photolibrary/Getty Images, 10 bottom right, 108; Terry Holland, 1, 12–13, 48 bottom, 118–119; © Terry Whittaker/Alamy, 10 bottom left; © Thierry GRUN/Alamy, 111 top left; © Tim Graham/Getty Images, 30; © Universal Photography/Alamy, 140; © UrbanHomestead.org, 54; © Veg Organic/Alamy, 70; © Wayne Hutchinson/Alamy, 100; © William Fawcett/fotoVoyager/iStockphoto.com: 83 right; © Wolfgang Kaehler/Getty Images, 139; © E. Carolina Sanchez Vazquez and Allan Aguirre Flores, Xome Arquitectos, 142; © Yuji Kotani/Taxi Japan/Getty Images, 146; © ZUMA Press Inc./Alamy, 14

© 2016 by Jono Neiger

Storey Publishing
210 MASS MoCA Way
North Adams, MA 01247
storey.com

Printed in China by R.R. Donnelley
10 9 8 7 6 5 4 3 2 1

Library of Congress
   Cataloging-in-Publication
   Data on file

FSC
www.fsc.org
MIX
Paper from
responsible sources
FSC™ C101537

CONTENTS

FOREWORD

Like many people, I came to permaculture via a circuitous route. In the early 1990s, my wife and I bought a house tucked into five wooded acres in the Cascade foothills east of Seattle. This being my first adult adventure in rural living, I spent long and happy hours in the public library paging through and dreaming over books on homesteading skills and country living. One of those books, a thick, black one, grabbed my attention. It had the word "permaculture" in the title, and I had no idea what that word meant. I pulled it off the shelf and browsed the table of contents. The subjects leapt out at me – design, nature's patterns, climate, trees, water, soils, earthworks, economics, community.

These were all topics I had been fascinated with my whole life, but I never saw how they fit together. I just thought I had a weirdly eclectic collection of interests. Wonderfully, permaculture assembled and related them in a way that made sense of my life and of the larger world. Those early permaculture books showed how all these seemingly disparate disciplines connect, mingle, synergize, and tie into an organic whole. The book you hold in your hands does this, too, but in a very different and much deeper way.

The first permaculture books were mostly theory, simply because there was so little permaculture actually on the ground and functioning. But now, permaculture landscapes can be found in more than 150 countries. North America abounds with dozens of mature, beautiful, and ecologically resilient examples of permaculture, and thousands of younger sites and permaculture-related businesses, schools, and nonprofits are popping up everywhere. Jono Neiger has explored many of them, designs them, and has built and lived in several of them, including his present home, Hickory Gardens, in Massachusetts.

This book is a bountiful harvest of the techniques, ideas, strategies, lessons, and wisdom embodied in the very best of permaculture design as practiced today. And although its main focus, like that of permaculture, is on working in ecologically sound ways with the landscape, in these pages you will also learn how permaculture can help meet every human and ecological need: food, water, shelter, energy, community, financial resources, health, and all the rest. Permaculture has burst out of the garden and into the community; *The Permaculture Promise* shows how that has been done and how you can apply permaculture almost everywhere, too.

Jono is uniquely suited to tell this important and richly faceted story. I met Jono at the very beginning of my permaculture career. He was the land steward and permaculture program director at Lost Valley Educational Center in Dexter, Oregon. I learned much from watching Jono's careful and sensitive guidance, care, and hard work heal Lost Valley's badly damaged, brutally clearcut 80-acre site. Now on the faculty at the Conway School and a principal in the firm Regenerative Design Group, Jono shares his wisdom with students and design clients. I'm delighted to see this book, as it brings me up to date on what Jono has learned and done in the years since he left the West Coast, and because it fills an important niche in the permaculture canon.

The Permaculture Promise does that in several important ways. It's chock-full of examples of permaculture used in landscape design, water conservation, waste treatment, carbon sequestering, community, and economics. The book also gives us clear, well-articulated definitions, concepts, and methods for applying the principles of permaculture, amply illustrated with dozens of captivating images. And best of all, it glows with an optimistic, can-do vision for our future. We are, as Jono writes, in an ecological and social mess, but we have powerful tools for fixing it. Many of them are covered in the pages that follow.

But this book is far more than an instruction manual. *The Permaculture Promise* is exactly that: a declaration, filled with living proof, of permaculture's present successes and barely tapped potential for building a resilient, abundance-filled culture that is a joy to live in.

– TOBY HEMENWAY,
author of *Gaia's Garden: A Guide to Home-Scale*
Permaculture **and** *The Permaculture City*

INTRODUCTION

HERE'S THE TOUGH NEWS: We live in a time of dramatic change and uncertainty. As human society reaches out to dominate every corner of the planet, ecosystems are fraying and natural resources are being consumed at an alarming rate. Climate instability and peak oil (the point at which the rate of global oil extraction peaks, thereafter dropping into decline) are inescapable realities. And all the while, the world's population continues to grow.

With such changes happening all around us, it's easy to feel like the world is out of control. It's easy to become cynical or to feel powerless. After all, each of us is just one person, a tiny speck in the seething mass of seven billion people on this planet. What could we, in our small way, ever hope to do to make things better?

The truth is, the only reality we *can* affect is our own — the immediate life we live each day. And for us, as humans, through all the great arc of our time on this planet, "reality" has been the sum of our relationships: to each other, to the world around us, and to that ineffable spark of life that innervates each of us. The human need for relationships rises in us and feeds us, physically, emotionally, and spiritually.

Yet modern life has a way of eating away at the relationships that support us. As we become more interconnected in the digital world, we seem to lose connections in the physical world; our social and emotional bonds become more fragmented, and we lose touch with our neighbors and community. As we move away from the land, we neglect the ecological and agricultural systems that feed and shelter us, and those systems become

more tenuous. As we begin to take for granted the energy, transportation, and distribution networks that support us, our understanding of these systems diminishes, and we begin to lose our sense of place as citizens and caretakers of a global world.

In this time of disconnect — of peak oil, climate chaos, population explosion, energy crisis, water shortages, mass extinctions, societal disruption — many people are wondering how to get from where we are to where we need to be in order to survive and *thrive* on this planet. The answer, I think, lies in building relationships. If we can build — or rebuild — connections to each other, to the land, and to the systems that support us, we can, perhaps, contribute to a growing worldwide web of interrelationships. That network, in turn, can become the foundation for a self-sustaining community that interweaves human endeavor with natural systems to support a resilient, prosperous future for all.

That's what this book is meant to provide: a perspective for navigating our future with grace and a long-term view, including some practical ideas for re-skilling, reconnecting, reengaging, and ultimately creating a more livable world for us all.

That is, in essence, the promise of permaculture.

IF WE CAN BUILD — OR REBUILD — connections to each other, to the land, and to the systems that support us, we can, perhaps, contribute to a growing worldwide web of interrelationships.

Restoring forests (such as this one in western New Zealand) sequesters carbon and creates habitat for native species.

A rice paddy in Vietnam provides dual crops: rice and fish.

Capturing rainwater for garden irrigation makes good use of a scarce resource.

WHAT DOES "PERMACULTURE" MEAN?

The term "permaculture" was originally coined by Bill Mollison and David Holmgren in their book *Permaculture One* (1978) as a contraction of "permanent" and "agriculture." Its etymology reflects the early concept of permaculture as primarily focused on agriculture. By the early 1980s, the definition had expanded in scope to broadly encompass the ways in which people can live on the land and in communities. Now, there are as many definitions of permaculture as there are permaculture designers — it's like a language, in the sense that it's constantly evolving as people participate in and contribute to it.

PERMACULTURE: A way for humans to consciously design systems that support ourselves – food production, energy, buildings, transportation, technology, even human relationships and financial systems – while acknowledging our roles as equal, co-creative members of natural ecosystems with the ability to regenerate our environment while we're providing for our own needs.

Permaculture focuses on mimicking the patterns and relationships found in nature. As we align our human society to match the way a whole, integrated ecosystem operates, we can't help but become healthy individuals within richly diverse communities. After all, the natural world around us is embedded with knowledge from millennia of trial and error. Unworkable and inefficient designs have largely been culled out, leaving only those that function as efficient, effective long-term systems.

DESIGN THAT CONSIDERS WHOLE SYSTEMS

Permaculture uses a whole-systems approach to design. A whole system, by definition, comprises all the parts and factors that contribute to the system's dynamic self-sufficiency and function. For example, when all the pieces of an ecosystem are functioning together and are mutually supporting, energy cycles efficiently and effectively; the living tissue of plants keeps carbon dioxide from being released into the atmosphere; water is collected and recharges local aquifers; and microbes, animals, and plants thrive and prosper.

Landscapes, communities, even organizations are like natural ecosystems — the more they're interconnected and diverse as a whole, the healthier and more resilient they become, and the less waste they generate. In fact, there really is no such thing as waste in nature; there are only resources moving from one place to another.

An efficient and well-functioning ecosystem cycles resources constantly. Yields from one part of the ecosystem provide for the needs of another part. In the same way, a brewery that yields spent grains can provide them to a mushroom cultivation or compost production operation, thus cycling the material and turning it into a benefit for the other business. The benefits are not just reduced waste and healthier food, air, water, and soil but also opportunities for both businesses to cut costs and support the local economy.

At home, whole-system design considers all the connections between the house and the landscape, as well as how water moves through the land and the ways in which it can be stored and used for irrigation, how people move from one part of the landscape to another, energy systems, food production zones, and areas intended for gatherings or for use as sanctuary. All these purposes interconnect, and when considered as a whole, they can be woven together to make an efficient, hospitable, serenely functioning system. If poorly planned, however, the different parts can feel disjointed and require more energy and maintenance.

A decision as simple as where to site a garden or plant fruit trees involves whole-system thinking. If placed farther away from the house, the garden may get less care and may be more prone to deer or rabbit browsing.

Just bringing a garden closer to the home and making it a central aspect of an outdoor living space can make it easier to tend and better utilized. Locating it in an area that has well-drained, fertile soil is clearly a good idea. Going a step further and incorporating the rainwater runoff from the roof and the graywater from the house (that is, wastewater from doing dishes and laundry) — not to mention a composting system for kitchen waste — would create a system that is even more resilient and self-sufficient.

Permaculture offers practical skills and solutions we can use to make the world more whole. It's a way for us to find our way back into a positive, life-affirming relationship with the planet that sustains us. And it starts small. Growing vegetables in your garden will feed you, lower your monthly food bill, and perhaps offer more nutrition than anything you would get in the grocery store, but it also gets you outside, in the garden, hearing, smelling, and seeing the daily activity and seasonal changes around you. When you think of gardening from this big-picture point of view — from a whole-system perspective — you come to realize just how much it feeds you, both physically and spiritually. It builds upon and nurtures our relationship and communion with the world, allowing us to provide for ourselves while acting as co-creative members of a larger system.

Growing tilapia and vegetables together in an indoor aquaponics system such as this is one way to make our food supply more resilient. It also "closes the waste loop"; the fish waste from the barrels fertilizes the growing vegetables.

A WAY FOR HUMANS TO BE MORE RESILIENT
IN A TIME OF CLIMATE INSTABILITY

Resilience is the ability to adapt to and bounce back from disturbances. Resilient people, communities, land, and ecosystems are able to continue functioning during and after difficult conditions. There is no way to know what the future holds for us, especially when it comes to climate change. In fact, "climate change" might be more fittingly called climate *chaos*; the climate is becoming very erratic, oscillating widely between hot and cold, dry and wet. According to scientists around the world, the fluctuations that we're experiencing now — including drought, flooding, extreme heat, extreme cold, and intense storm cycles — are a harbinger of the future. In order to get through the difficult times to come, we need to cultivate the ability to adapt to change, to anticipate and account for these fluctuations, and to design resilience into our systems.

Like a building that has been designed and built to move and flex with hurricane winds, a garden that grows a diversity of crops can handle a few pests and still yield well. Better yet, if the plants are healthy and strong and the garden includes plants that actually repel pests, perhaps the pests just pass by. On a larger scale, an uncertain future will necessitate that we rethink our current industrial food-production system and rebuild our local systems of processing and distributing food, energy, and materials. In this way, we support diversity in our infrastructure and resources while allowing smaller regional systems to adapt to local conditions and requirements. At the same time, we can rebuild the social capital that binds together and supports our communities.

A POSITIVIST APPROACH TO THE
CHALLENGES OF OUR TIMES

Permaculture is "positivist," which means that we can acknowledge the difficult realities of the world, but it is important to focus on what we *can* do, not what we *can't* do. We keep our eye on what is good and possible, avoiding the cynicism and negativity that are so easy to slip into. This positivist mind-set provides us the incentive to make a difference rather than despair about those things we can't control. Can we change our lives and begin to live as part of the ecosystem around us? Can every action we take, every building we build, every crop we grow, and every human relationship

we develop help make the world a better place? This is the vision of permaculture: creating a world that we engage with and *improve* as we live, work, and play.

As a keystone species — one that has taken control over vast areas of the planet — we need to become better caretakers, to see ourselves as *part* of nature, not above it. By examining our lives through a permaculture lens, in everything from how we grow food and create built environments to how we care for each other, we can help make the world a better place.

Unlike traditional farms that till up the soil each season, "no-till" farms such as this one plant directly into the previous year's crop residue without turning the soil. The no-till method maintains soil structure, prevents erosion, and decreases the amount of carbon released into the atmosphere.

A WAY TO REGENERATE THE EARTH
WHILE WE PROVIDE FOR OURSELVES

Living systems are regenerative; that is, they renew themselves. The materials and energy that drive all the growth, interactions, production, and decay in a living system are cycled through that system again and again, whether quickly (over the course of a growing season) or slowly (through the geologic ages of the planet). But when those materials and energy are subtracted from a living system — like fossil fuels being extracted, or nutrients being washed away, or a major species being eliminated — the system's ability to regenerate suffers a blow. With enough damage, the system loses the ability to regenerate entirely, and a cascade of environmental "crashes" follows.

We're facing a cascade of environmental crashes right now. Living systems across the planet are failing: desertification (the creation of new deserts from previously fertile land) is on the rise, dead zones in the oceans are growing, topsoil is increasingly absent or depleted, major species everywhere are lapsing into extinction. But permaculture offers one simple principle to combat the entire onslaught of disaster: Support nature's ability to regenerate.

Happily for us, supporting nature's ability to regenerate itself is also one of the best ways to make the land more productive, stable, and generally conducive to supporting humankind. When we actively participate in and help create regenerative systems (or help repair damaged ones), we can provide for our needs in ways that restore the earth and build its capacity to hold more life. Creating and supporting regenerative systems means moving beyond "sustainability" and actually giving back to the ecosystem around us as we take care of our own needs.

Communities need regenerative work as well. People need connection, to each other and to nature. Reinvesting in communities and rebuilding social networks and social capital bring multiple benefits, like improved education, better health through cleaner water and air, and strengthened connections between elder and younger generations so that, over time, a community holds together and renews itself. In many ways, community regeneration is harder than planting a garden, but the yields can be greater.

AN ETHICAL APPROACH TO LIVING

Ethics are guideposts for what you might call "right action." Over time, societies develop these guideposts to help people live in alignment with their values. Permaculture has its own set of ethics, and they are three-fold: earth care, people care, and fair share. The first two are probably self-explanatory. The principle of "fair share" simply refers to recognizing the limits of what we need and not consuming more than is necessary, so that we can allow the excess to flow to others who are in need.

These ethics help us regulate our own self-interest and identify when our actions have strayed from the "right" path. For example, if a farm's practices result in contamination of nearby waterways with animal manure, if the farm's workers can't afford housing or food, or if the farm's surplus isn't being shared, it becomes clear that there is a disconnect at some point in the system, because permaculture ethics aren't being followed.

CARE FOR THE EARTH:

Don't allow nutrient runoff to pollute local waterways.

CARE FOR PEOPLE:

Pay a living wage so that people can take care of their families' needs.

FAIR SHARE:

Allow neighbors in need or community-supported agriculture (CSA) members to harvest surplus or "seconds" crops.

With these three ethical principles as a foundation for planning and design, permaculture can guide us in our work so that we are always aimed toward right action.

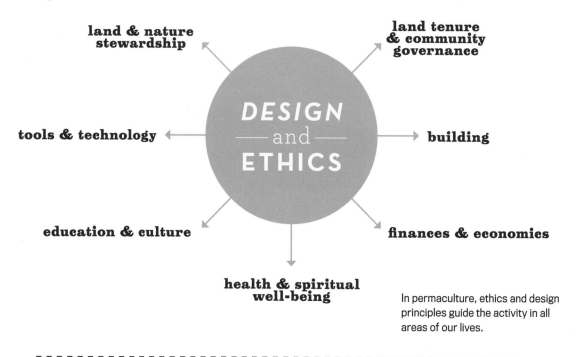

land & nature stewardship

land tenure & community governance

tools & technology

DESIGN —and— ETHICS

building

education & culture

finances & economics

health & spiritual well-being

In permaculture, ethics and design principles guide the activity in all areas of our lives.

"A man is ethical only when life, as such, is sacred to him, that of plants and animals as that of his fellow men, and when he devotes himself helpfully to all life that is in need of help."

~~~~~~~~

—Albert Schweitzer

PREPARING FOR THE FUTURE BY LEARNING FROM THE PAST

Permaculture is really nothing new. Many of the principles and practices of permaculture have been known and used by peoples in different settings, climates, and cultures around the world for millennia. For their own preservation, indigenous societies historically developed practices to live more in tune with nature and within the limits of their local ecosystem.

The indigenous philosophy is one that proponents of permaculture now strive toward. But most of us have come far from our own indigenous roots; as newcomers to the land, we will need time to "become native" – time to gain experience on the land and learn to work within its patterns and limits.

The Woodbine Ecology Center in Littleton, Colorado, has laid out five principles defining indigenous permaculture. They are:

1. The recollection and recognition of, and respect for, indigenous contributions

2. Traditional ecological knowledge that is specific to place

3. Decolonization of our minds, our language, our work, and our communities

4. Being and becoming native to your place, with protracted and thoughtful observation

5. Eco-cultural preservation and restoration – that is, the preservation and restoration of natural places in concert with the preservation and restoration of the indigenous cultures of those places

As an example, consider the traditional Vietnamese "VAC" – an acronym for the Vietnamese words for garden, fishpond, and animal sheds. This integrated home garden exemplifies the manner in which many native peoples came to actively manage their environment to benefit local ecosystems while providing for their own needs. In a VAC, waste from animals feeds fish, the pond water fertilizes and irrigates the gardens, and overlapping layers of productive vegetation yield vegetables, fruit, and herbs. A VAC can provide for the majority of a family's needs, and the surplus sold at local markets might provide substantial income. At the same time, a VAC is home to an incredible diversity of flora and fauna and efficiently cycles energy, water, and organic matter in a mostly closed-loop system. VAC gardens can still be found throughout Vietnam today. Some of these gardens are three hundred years old and still producing food! Grounded in ecological principles established over centuries of experience, they serve as inspiration for those of us who, looking ahead to an uncertain future, hope to build upon the past.

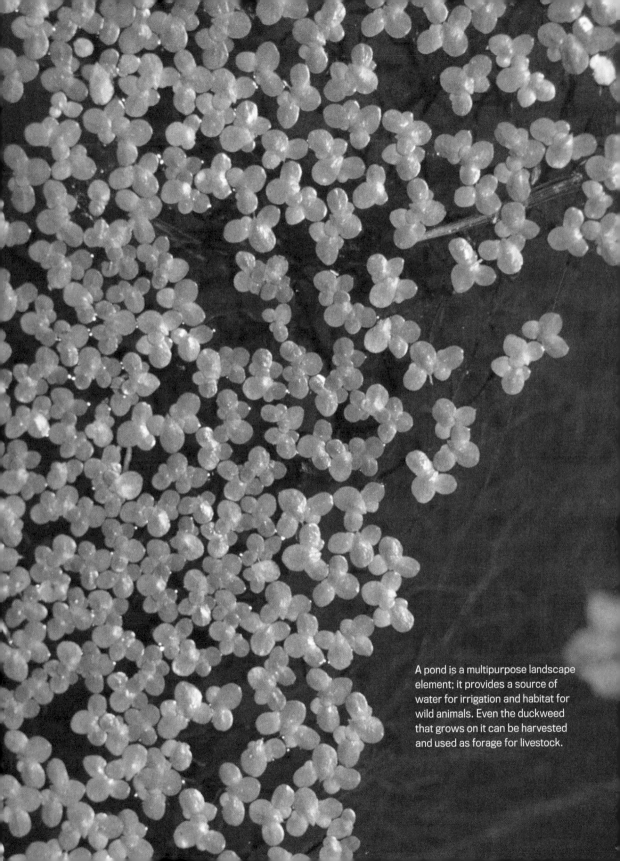

A pond is a multipurpose landscape element; it provides a source of water for irrigation and habitat for wild animals. Even the duckweed that grows on it can be harvested and used as forage for livestock.

PERMACULTURE
CAN

(1)

Turn Deserts into FARMLAND

We often assume that human impact on the environment must necessarily be destructive. At best, we think, we can make an effort to be "sustainable" — something the United Nations has defined as "development that meets the needs of the present without compromising the ability of future generations to meet their own needs."

But what if we could do better than that? On a planet where humans dominate so much of the land and resource base, what if we could do better than simply not compromising the future? Creating regenerative systems — those that revitalize, restore, and rebuild — moves us beyond sustainability. Regenerative design builds up life; it is expansive, additive, affirmative. It enables us to repair the damage that humans have wrought on the environment.

All living systems are naturally regenerative; they renew and rebuild their materials and energy over time. A forest grows more complex as the years and decades pass, adding species, biomass, and interconnections. Disturbance — brought by a storm that blows down limbs, or loggers who harvest trees, for example — will knock it back, but the regrowth process inevitably begins again. Trees will heal their wounds, and saplings will fill in open land. Soils are another example of regeneration; they grow in depth and fertility over time, given the right conditions.

Some environments are more fragile than others, however. In some parts of the world, human interaction has degraded landscapes to the point that they have lost their ability to regenerate without some assistance. In vast swaths of dryland ecosystems, for example, human interactions have promoted desertification, with almost complete loss of vegetation and animal life. Drought and climate shifts also play a role, of course; it's harder for a natural area to regenerate if there's no water.

The Loess Plateau in northwest China suffered from centuries of overuse and had one of the highest soil erosion rates in the world before it was restored (above in 1995). A large-scale restoration project not only revitalized the land (shown below in 2009) but reduced sediment runoff into streams and brought millions of people out of poverty.

Legumes (members of the pea family) like this pigeon pea tree (*Cajanus cajan*) help enrich the soil by "fixing" nitrogen from the atmosphere, which makes it available for other plants to utilize.

REGENERATE, RATHER THAN DESTROY

Although humans have contributed to the degradation of these lands, we can also take part in healing them. By using contour swales to help capture the small amount of water that is available and by planting crops that return nutrients to the soil (such as leguminous trees, which fix nitrogen) and serve as windbreaks to slow erosion, we can reverse the process of desertification.

In some areas of Africa, farmer-managed natural regeneration (FMNR) has been successful not only in preventing and reversing desertification but also in restoring degraded cropland, grazing land, and forests. In many dryland areas, this "underground forest" of resprouting trees had been neglected and even removed for cropping and tree-planting projects. Under the FMNR system, farmers now encourage and manage the natural regrowth of woody plants and support the cycle of healing. The trees and shrubs are managed carefully with traditional coppice and pollarding practices — pruning methods that produce harvestable wood without removing the tree. By supporting the lands' impetus to regenerate, farmers receive benefits such as firewood, fencing, crop shade, soil improvement, and fodder for livestock. Farmer-managed natural regeneration is a low-cost, low-input, large-scale regenerative approach led by locally knowledgeable people.

When we are in sync with life's processes and working with natural systems, we can build up the soil, the health of the land, and the ability of communities to support themselves.

Rather than degrading the land, and thereby fraying our connections in the web of life, we can reinvest in forms of regenerative caretaking, so that the land can support a diversity of life, both human and not.

REGENERATING COMMUNITY

JUST AS WE CAN HELP foster regenerative systems in the physical environment, we can design systems that build up capital in other capacities, whether it's social, material, financial, cultural, intellectual, experiential, living, or spiritual. Thinking of capital in these many forms moves us away from the idea that only money can help build resilience and strength in communities. In each of these areas, we have resources to rely on, build, and expand. How we act can either restore or degrade any of these forms of capital. For instance, places without agriculture and resilient food systems are low in living capital: plant, animal, soil, and water resources. Communities can build living capital by developing community gardens, establishing plant exchanges and seed libraries, and connecting consumers to regional food growers through farmers' markets and community-supported agriculture farms. They can protect agricultural land that is under threat of development and help connect small farmers with affordable land.

How do we know if an activity or system is regenerative? Any system, whether it deals with environmental or some other form of capital, is not regenerative if it passes a problem along to some other group or some other system, or if conditions are degrading rather than improving.

There are simple ways to gauge this. Are people employed and able to support their families? Does the community benefit from the activity? Are youth mentored and elders respected? Does money cycle many times through the community before leaving? Can people afford to live on the land? Do gardens and farms support the community with locally produced food? Is the soil improving? Regenerative practices make life better for everyone involved and help communities move beyond sustainability to a healthy, vital future state.

Shopping locally at a farmers' market is one way to contribute to and regenerate your community.

Village Homes: Designing an Integrated Community

Walkable communities like Village Homes increase the amount of interaction residents have with each other and with the landscape they share.

IN MOST SUBURBAN developments in the United States, little consideration is given to community life. Common space is limited. Homes are designed for individual families, each with its own yard, garage, and lawnmower. Houses are set as far apart from each other as possible and separated by driveways, for privacy. Stormwater is channeled off-site as quickly as possible (for more about that, see page 104), and productive parts of the landscape are separated from decorative parts. Everyone has separate lives, and the connections between people, young and old, are lost. You might live next door to someone for years and never meet them.

But there are better ways to design a subdivision and, in the process, create a healthy, vibrant community. Village Homes in Davis, California, is one such example.

This 70-acre subdivision was designed and built in the 1970s and 1980s by Michael and Judy Corbett. Many of the design concepts are drawn from earlier "greenbelt communities" — residential areas that are organized around open space. As an example of integrated design at a community scale, it is unique. Bill Mollison, one of the pioneers of permaculture, called Village Homes "the village of the future."

The community consists of 225 homes and 20 apartments, each uniquely designed, many of them with active or passive solar. The houses are clustered on the property, to provide significant open space for the community, and are oriented along east–west running roadways, to allow for solar electric generation at each home. More than a third of the land in this community is set aside in greenbelts, orchards, vineyards, gardens, and parks. The land is managed mostly organically, and residents can freely harvest from the edible landscaping. An almond orchard along one edge of the land is cultivated for a commercial crop, with the revenues helping to pay for land maintenance. A central area has a large community gathering space, which also acts as a stormwater collection basin during storms.

The streets are narrow and tree lined, and cars are parked in a lot that is separate from the living and community spaces. The houses are connected by bike and walking paths. The neighborhood is also well connected by bike paths to Davis; many of the residents commute into town or to the nearby university, UC Davis.

In Village Homes, we see the weaving together of homes, a few businesses, community, food, open space, stormwater management, foot-powered transportation, and energy conservation and production. The elements are combined into an integrated whole that allows residents to live as co-creative members of the community. The focus on sustainability, with solar power, green design, open space, community gardens, and water runoff management, allows Village Homes to be less destructive and more regenerative than almost any suburban development on record. Many more such developments can and should be designed and constructed to replace our aging suburban infrastructure and begin to transform the way we live today.

2
Create
SELF-FERTILE SOIL

Many people garden using bags of compost and fertilizer from the garden center. Buying in fertility may be a stopgap measure for getting your home garden or farm up and running, but it's not a long-term solution. It's a bit like being on a treadmill. Every new growing season means another trip to the garden center and another bag of fertility brought in from off-site.

Comfrey

A better strategy for maintaining the fertility of our soil is to use the plants and animals already working there. This is self-renewing fertility, and it happens naturally in wild places. Plants reach deep into the soil layers to access and concentrate nutrients, which then become available to plants and soil microorganisms when the plants die and decompose. Or animals consume the plants, such as pasture grasses, and leave behind (as manure) what they cannot absorb. On marginal land, the rich manure from these grazers is especially important for improving depleted soil.

We can replicate this natural system of self-renewing fertility in human-created landscapes. There are several simple but key factors:

PLANT "DYNAMIC ACCUMULATORS." These plants excel at gathering nutrients and accumulating them in their tissues. Once these dynamic accumulators decay, whether naturally or via composting, those nutrients are released into the soil and become available to other plants. Dynamic accumulators can do much of the work of building soil. We don't know everything about how these plants work and where in the plants most of the nutrients accumulate. Certainly much of it happens in their roots, which are in the topsoil layers. Some scientists also believe that deep-rooted plants can pick up nutrients from deeper soil layers, but this theory is still being researched. Comfrey is an excellent example of a deep-rooting herbaceous plant that accumulates nutrients, including potassium, phosphorus, calcium, copper, iron, and magnesium. Watercress, a floating aquatic perennial, accumulates many nutrients. Some other excellent dynamic accumulators are stinging nettle, black locust, dandelion, sorrel, dock, black walnut, and shagbark hickory.

GROW PLANTS THAT ADD NITROGEN TO THE SOIL. Nitrogen is a limiting nutrient in many soils, meaning that it is often in short supply and may be the first nutrient to limit a plant's growth. And since it is water soluble, nitrogen can easily be leached out of soil if it isn't bound to organic matter and replenished. Nitrogen-fixing plants, primarily those from the bean and pea family, have a symbiotic relationship with a particular kind of bacteria that helps them convert nitrogen from the air into a form that is usable by plants.

Plants such as vetch, clover, and alfalfa can convert as much as 250 pounds of nitrogen per acre. However, the nitrogen is concentrated in nodules on their roots and goes quickly into seeds as they ripen; in fact, as much as 70 percent of the captured nitrogen can be tied up in the seeds. To make this reserve of nitrogen available to surrounding crops, the plants must be cut back before they produce seeds. The stress from being cut back causes the plants to shed roots (and thus nitrogen) into the soil.

CREATE A FERTILITY BANK. This is a cache of woody or herbaceous perennial plants that are grown specifically so their new growth (fertility) can be harvested and cycled back into the garden, orchard, or field. Plants are cut back yearly (or several times per year), and the trimmings are either composted or applied directly to the soil. In this way, the organic matter built up in the trimmings can be incorporated into the soil.

Plants like comfrey excel at gathering nutrients, which are released into the soil when they decay.

With woody plants, this process is called "coppicing"; with herbaceous perennials, it is called "cut-and-come-again" or "chop-and-drop." Perennial plants that form the base of a fertility bank will regrow from the same root systems, so they do not need to be replanted year after year.

In one fertility bank my design partners and I planted, we used switchgrass, bush clover, creeping comfrey, and Russian comfrey. An autumn olive self-sowed there, and its high-nitrogen branches and leaves are coppiced. Compost bins nearby can be used to process the vegetation into compost, or it can be used as mulch directly in growing beds.

KEEP SOIL MULCHED OR COVERED WITH PLANTS. Bare soil is an invitation for erosion by wind or water, overheating, and evaporation, which can lead to leaching of nutrients and loss of soil fertility. Also, weed seeds that are exposed by tillage quickly establish in bare soil. Soil should be covered, ideally with thick vegetation or mulch, at all times. Even a thin layer of mulch over soil that has just been worked will reduce the damage to soil structure, allow beneficial fungi and microbes to stay active, and protect soil's ability to maintain fertility.

The fertility bank at Wildside Gardens features switchgrass, bush clover, and comfrey. These plants are cut back hard each year, and the prunings are used as high-nitrogen organic material for the production garden.

A better strategy
for maintaining the fertility
of our soil is to use
the plants and animals
already working there.

3

TURN WASTE *into* FOOD

Everything a natural ecosystem produces is cycled, from soil, air, or water through plants and animals and then back to soil, air, or water again. When soil, water, sun, and seed conspire to produce a tree, it grows and grows, storing organic matter and nutrients in its body, housing animals and insects in its branches, providing shade and shelter for surrounding plants.

When the tree dies, woodpeckers break it up digging for insects, insects burrow through the decaying wood, and as the wood is broken down by fungi, it returns to the earth to become soil. The soil grows another tree. Around and around it goes. There is no waste, only cycles of reuse. Everything is food for the next part of the cycle.

REJOIN THE CYCLE. Somewhere along the timeline of human agriculture, we began breaking the cycles of reuse into linear chains, and seeing the materials at the end of the chains as waste. Manure from farm livestock was no longer cycled back onto farm fields and thus became a waste product to be disposed of. The fields then needed fertilizer, which was made in factories and became an additional expense for the farmer. Excess fertilizer ran off the fields and caused pollution in nearby waterways and groundwater. With cycles of reuse broken and waste products rampant, problems began to multiply.

When we focus on turning "waste" into food, we are resurrecting natural cycles that have been disrupted. Following the tenets of permaculture, we can provide for our own needs while participating in and supporting the great cycles that are so essential to natural ecosystems. We can work with this cycling and do the work of growing soil, and thus the plants and animals we rely on.

COMPOST EVERYTHING. One of the easiest ways to convert waste to food is by composting, which captures the organic matter in food scraps, prunings, lawn clippings, and other plant debris. Organic matter is an essential ingredient for building rich soil, which in turn produces the crops we consume. In almost all circumstances, more organic matter is called for. Is your soil too clayey? Add compost. Too sandy? Add compost. Just plain depleted? Add compost.

Be on the lookout for other sources of organic matter that might be flowing through your community as "waste." Arborists and companies that clear trees away from power lines often have truckloads of wood chips looking for a place to go. Fallen leaves (often bagged up by the side of the road), pine needles, and woody brush are also usually available for the taking. They can be added to your compost or chopped and used as a mulch that will decay in place, providing organic matter over the long term, while at the same time helping to increase soil life, reduce evaporation of water, and cool the ground in summer.

UTILIZE HUMANURE. In most homes, the fertility of our own human waste is taken away, whether to a municipal waste treatment plant or into a backyard septic system. Although it's an uncommon practice in our culture, this "humanure" can be a fertility source instead of a pollution problem. Simple systems to compost and cycle humanure, urine, and graywater (household water, such as that from the laundry or bath, that does not contain serious contaminants) can save large amounts of water and energy and contribute to the self-renewing fertility of our land. Projects to demonstrate and document the benefits of cycling human waste are important. The Rich Earth Institute in Brattleboro, Vermont, demonstrates the use of urine as an agricultural amendment, closing the loop of pollution caused by the high nitrogen in urine. Instead of polluting water resources, it's improving the production on croplands.

ENRICHING SOIL WITH BURIED WOOD WASTE

WHEN I FIRST was learning about permaculture, I was tending an orchard in the Pacific Northwest that had been planted in challenging clay soils. The extreme rains of winter gave way to scorchingly hot, dry summers; the soil was either too wet or cracking dry. To assist the fruit trees, we dug trenches in the clay soil — 4 feet wide, 2 feet deep — leading across the orchard and toward a small stream. We filled the trenches with woody debris and logs and backfilled them with compost and soil. When the winter rains came, the trenches helped drain the excess water from the adjacent clay soils, funneling it to the stream. The woody debris soaked up some of that water and slowly released it through the summer. The decaying wood in the trenches also created a nutrient reservoir for the trees to root into. Over time, the trenches, woody debris, and plentiful mulching began to shift the balance toward a more even, fertile soil.

This kind of system is called *hügelkultur* – roughly, "mound culture" in German. Wood is stacked on the ground (or in shallow trenches), and a layer of soil or compost is shoveled over the top. The mound becomes a raised garden bed, and the plants growing there can take advantage of the slow decomposition of the wood, which holds moisture and slowly releases nutrients. This sort of setup takes "waste" woody materials and uses them to build the soil, helping us provide for ourselves while regenerating soil fertility.

Woody debris that is stacked and buried, in a system called *hügelkultur*, creates a thriving microbial habitat that enriches the soil and feeds plants. The wood waste also absorbs and retains moisture for plants to draw on during droughty periods.

4

SPREAD
the WEALTH

There are now more than seven billion people on the planet, and our numbers increase every day. Many of these people are struggling to survive, while a small percentage thrives. What would it look like if we created a society that met everyone's basic needs?

The founding fathers of the United States said (I'm paraphrasing) that all people deserve "life, liberty, and the chance to grow kale." Similarly, permaculture calls for "people care" — in short, that everyone should be able to take care of their needs for food, water, shelter, and personal connection.

However, centuries of land and power consolidation have left many people around the world without access to land, with no way to sustain themselves and no basic skills to help them meet their everyday needs, such as collecting water and growing food. The vast wealth of developed societies is out of reach for most people.

Just as we can reinvigorate frayed ecosystems by layering the soil with organic matter and creating habitat for pollinators, we can regenerate degraded communities by sharing resources — physical, mental, and emotional. We can limit our own consumption and distribute the surplus more evenly.

Economies and other systems are based on exchanges and the flow of "goods," whether they be currency or food. Surplus results when an economy or a community reaches abundance. Too often in human societies, though, the surplus remains with a select few instead of being shared with those in need. This parsimony — keeping for ourselves, rather than sharing with those who need it — goes against the principles of nature, however. After all, earth's systems are designed to create and move surplus, whether it is in grain, tomatoes, labor, or love. When animals reproduce, the offspring often move on to populate new areas. Some plants produce thousands of seeds, which are carried on wind, water, and wing to spread far and wide. Wind, water, heat — all forms of nature's energy concentrate, build, spread, and dissipate. Onward and outward, natural systems distribute surplus to support balance and stability.

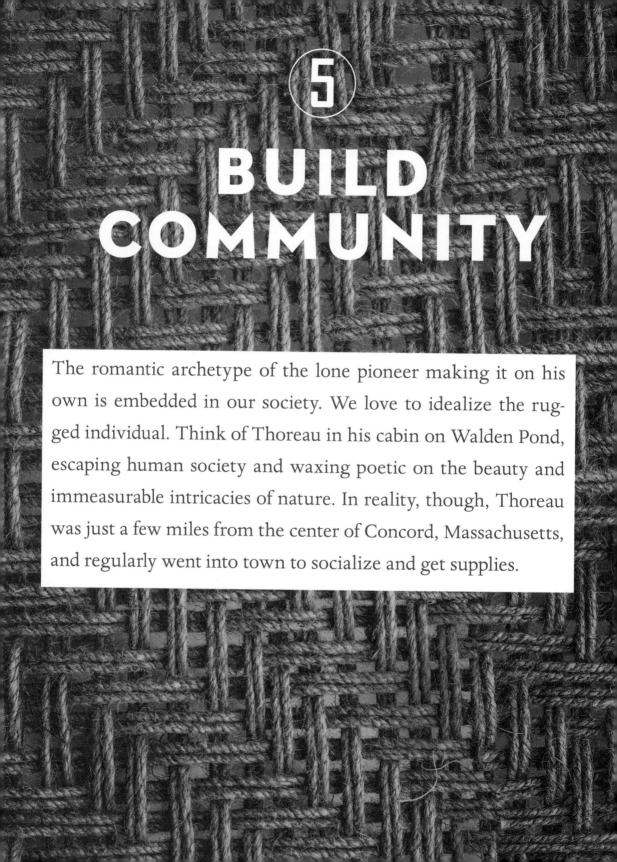

5

BUILD COMMUNITY

The romantic archetype of the lone pioneer making it on his own is embedded in our society. We love to idealize the rugged individual. Think of Thoreau in his cabin on Walden Pond, escaping human society and waxing poetic on the beauty and immeasurable intricacies of nature. In reality, though, Thoreau was just a few miles from the center of Concord, Massachusetts, and regularly went into town to socialize and get supplies.

Just like Thoreau, we are inescapably a part of society, and we need the connections and mutual support that come from living with others. A strong community is a source of resilience in the face of an uncertain future. In even a relatively simple world, the skills you need to live on your own and be truly self-sufficient are vast: growing food (vegetables, fruits, grains), raising animals, preserving food, building shelter, making medicine. Add in the skills of a more modern world — water filtration, energy system management, engine repair, plumbing, wiring, electronics repair, and so on — and most of us will simply throw up our hands in despair. We may be adept in some of these fields, but it is the rare, extraordinary person who can claim to grasp the fundamentals of all of them. And that is why we need each other.

The lone pioneer is really just a myth. We *need* our neighbors. We need lots of people within the community who can each play a role — to grow vegetables, teach the children, represent us in government, care for elders, build our homes, repair our infrastructure, and on and on. Start by getting to know your neighbors and learning what their skills and passions are. As a community in search of stability and resilience, figure out which pieces of the puzzle are missing — which skills or resources your community lacks — and make a plan for filling in those holes in your community base.

As with other permaculture facets, it's the interconnections between the parts that build the complete picture. In a healthy, whole community that functions with resilience, you can contribute your skills, do what you are passionate about, and know that others will be there doing their part, too.

A strong community is a source of resilience in the face of an uncertain future. . . . The lone pioneer is really just a myth. We need our neighbors.

WAYS TO BUILD COMMUNITY

You can build community in many ways. Do something slow and small to start, then let it grow. Bring others together, share ideas, and soon the momentum will build. Here are a few ideas:

HAVE FUN TOGETHER.

Organize community events, such as a seed swap, tool swap, potluck, reading group, or speaker series.

LEARN ABOUT WHERE YOU LIVE.

Get to know the history, the culture, the stories.

MAP YOUR COMMUNITY RESOURCES.

Put the map online or share it with others somehow.

CREATE COMMUNITY SPACES.

Plant community gardens, set aside land for a playground, or build a community center.

SHARE SKILLS.

Bring people together to share a skill, such as saving seeds, making compost, sharpening tools, or making bread.

VOLUNTEER.

Offer to help a neighbor with child care or to bring dinner to families experiencing hardship.

ASK QUESTIONS OF YOUR ELDERS.

Teach children about the place where they live and the natural world around them.

DESIGN TOGETHER.

Host a "permablitz" (an informal gathering of friends or volunteers to take on a project, like establishing a garden), or a design charrette (a more formal, intensive, interactive group design session).

HELP INDIVIDUALS & COMMUNITIES RE-SKILL

Our culture has taken skills specialization to an extreme. As an example, farmers make up 3 percent of the population and grow food for all the rest of us, using industrial practices such as large diesel-powered equipment, chemical fertilizers, and pesticides.

If asked to grow food, most of the population would be unable to — and we'd starve before we could figure it out. Similar statistics hold true in just about every other field. We are a nation of specialists, and with that specialization comes lack of resilience and the failure of our basic systems if supply chains become broken, technology fails, or people with specialized skills aren't nearby.

Unfortunately, most of us are completely disconnected from the sources of our food, clothing, water, and energy. The skills we need to transition to a low-fossil-fuel society aren't taught in schools and often aren't kept alive within our communities. Building shelters, managing land and water resources, making clothing and shoes, making medicine, growing food and fiber; these skills are learned over long periods of time and are often specific to location and culture.

We don't each need to be entirely self-sufficient, but familiarity with basic life skills is important for long-term resilience. A higher level of self-sufficiency builds self-respect and self-reliance, reduces dependence on faraway sources of food or energy, and builds connections within the local community. When we are confident that we can care for ourselves and our communities, we are stronger and more secure.

7

MAKE
Wetlands & River Edges
PRODUCTIVE

Historically, low-lying wet areas (wetlands, seeps, swamps, boggy meadows) have been thought of as wasteland, unsuitable for productive use, and have been drained so that they can be put to use for development or agriculture. These environments are disappearing worldwide, from the peatlands of Southeast Asia to the coastal mangrove forests of North America, West Africa, and Indonesia. Globally, we have lost half our wetlands since 1900, and the remaining ones are at risk from development, pollution, dams, and climate change.

Arcata Marsh and Wildlife Sanctuary on Humboldt Bay in Northern California is a reclaimed wetland that treats the town's waste water, serves as wildlife habitat, and is even a recreation destination.

47

In our battle against nature, though, we have lost sight of the important roles that edges and wetlands play. They offer critical habitat for a variety of species. They serve as floodwater storage and shoreline stabilization, protecting areas that are inhabited by people, and can buffer the effects of storms and rising sea levels. They filter runoff water from storms, keeping rivers and oceans cleaner and reducing sediment buildup. Wetlands are also important carbon sinks and can play a large part in mitigating climate change.

Besides playing important roles in natural ecosystems, wetlands and river edges can support many crops. Though aquaculture and wetlands agriculture can be problematic if nutrients from traditional land-based agriculture escape into the waterways, the potential for growing food here is huge. Fish can be raised in rice paddies, and crawfish with wasabi plants. Some species, like duckweed, can be grown for fodder and have biofuel potential. Poplar and willow along the edges can provide biomass, medicine, and biochar (a kind of charcoal that can be used for enriching garden soils). Blueberries and elderberries, among other fruit-bearing shrubs, love the water's-edge conditions. Find the right crops to utilize these edges and they will thrive and produce abundantly.

But agriculture is not the only productive use for wetlands. On Humboldt Bay in California, the Arcata Marsh and Wildlife Sanctuary is a former wetland turned industrial area that was reclaimed and converted back into wetland. It's now utilized for treating the town's wastewater, an innovative practice that has earned it many accolades. It also serves as a wildlife sanctuary and recreation destination. (For more on Arcata, see page 50.)

Wildside Gardens (see the profile on page 112) utilizes a wet meadow for growing rice in a paddy.

CHINAMPAS: GROWING ALONG THE EDGES

The practice of using and expanding edges of waterways (ponds and rivers) for food production began in areas of Central America and Southeast Asia long before European colonization. The Aztecs built the city of Tenochtitlan on an island in Lake Texcoco and surrounded it with constructed "floating gardens" or artificial islands, called *chinampas*, made with the rich shallow lake muck. They farmed intensively on these gardens, cultivating vegetables, herbs, fruit, medicinal plants, weaving and biomass materials, and animal fodder crops. With this method of agriculture, they were able to produce much of the food for Tenochtitlan while also helping to stabilize the island's edges and creating habitat that would support fish and waterfowl.

Arcata Marsh and Wildlife Sanctuary

IN 1974, A POLICY WAS enacted in California that restricted the discharge of wastewater into bays and estuaries unless that wastewater could be shown to enhance the receiving waters. Around that same time, on Humboldt Bay, in Northern California, the city of Arcata was faced with needing to upgrade its waste-treatment system. Engineers recommended building a regional treatment plant that would send the waste pipe from the treatment plant far out from the bay, where it would be diluted by ocean waters and have less risk of polluting the fragile bay waters.

The people of Arcata demanded a better solution. As a result, a large sewage-treatment marsh was constructed beginning in 1981 on derelict land along the bay. Studies showed that the wetland ecosystems could serve as a natural filter for wastewater and, where the treated wastewater joins with seawater, that young salmon and trout can be raised. In this way, the wastewater *enhances* the ecology of the wetlands and bay, serving as a resource for environmental diversity, rather than a source of pollution.

Now at 307 acres, the treatment marshes are part of a wildlife sanctuary that acts as an important stopover point for migratory birds. There are freshwater and saltwater marshes, ponds, estuaries, and other wetlands ecosystems brimming with diverse life. Meanwhile, the city of Arcata saves money on infrastructure and energy. The Arcata Marsh and Wildlife Sanctuary is an example of how alternative solutions can arise when we examine our goals and evaluate them in terms of how they're connected to a larger system. A standard waste-treatment plant would have performed the function of treating waste, but without the added benefits of a treatment marsh and wildlife sanctuary.

The Arcata Marsh and Wildlife Sanctuary is a diverse, multiuse "edge habitat" that filters wastewater and hosts more than 300 species of birds.

8

CREATE *More* LIVABLE CITIES

Using permeable pavement instead of solid blacktop or cement is one way to make cities more livable. It allows rainwater to infiltrate into the soil, keeping it on-site instead of flowing into local waterways.

More than half of the world's population now lives in urban areas — intensively built environments that rely on resources coming in from faraway places. City dwellers depend on food trucked in from the countryside or beyond, building materials and commercial goods shipped in from around the world, a system that pipes in water from reservoirs, and another system of piping to take away waste. It's a linear system of materials coming in, being consumed, and waste being taken away. Such a system is wasteful and energy intensive.

In addition to all this, open land for growing food within cities is very limited and often contaminated. Much of the land is paved over. This not only decreases land for growing food, it also causes stormwater to quickly flow off the land instead of into the soil, for plants to utilize or to recharge groundwater. As a result, stormwater flows into sewer pipes and nearby waterways, carrying sediment and other contaminants with it.

It's not difficult to see how fragile this kind of environment is. Take away any of the human-constructed systems we've set up, and the city isn't able to function properly. Considering the vast numbers of people who live there, finding ways to make cities more resilient is imperative.

THE URBAN ECOSYSTEM

Healthy ecosystems produce no waste, cycle nutrients, and support a variety of species. By that definition, you might say that the urban environment is an example of an *unhealthy* ecosystem. In many cities, waste is taken "away," potential nutrients from food waste and human waste are not utilized for growing food, and by far the dominant species is *Homo sapiens*.

If you take a closer look, though, you might see a different story. Urban areas actually mimic what ecologists call "early succession" environments. They're characterized by disturbed land with lots of "edge zones," where a wide range of tough plants and shrubs take root and provide habitat for a diversity of birds and small mammals. In cities you can find these "weedy" environments everywhere; even unassuming areas like the ground along chain-link fences contain plants struggling to become established. And though you might not realize it, there are actually many species of wildlife in urban places. Birds are often prolific, if not always as diverse as they might be, and small creatures can thrive, often coming out at night or moving along the edges and through unmanaged pockets of land. Thinking of the city as a "natural" place waiting to be revitalized can help us find appropriate solutions to the problems that plague urban environments.

The gardeners at the Path to Freedom urban homestead grow up to 6,000 pounds of produce in just under 4,000 square feet of space.

BIG PROBLEMS, SMALL SOLUTIONS

Even though the problems may seem overwhelming, instituting small, strategic changes to restore the balance in urban systems and encourage more self-sufficiency can make a big difference.

CREATE SMALL-SCALE, INTENSIVE GARDENS. Cities do not tend to produce much food for themselves because open land is limited. But small spaces can be very productive — in fact, the amount of food production per square foot can be much higher in small plots than in larger areas. Think of it this way: If you have a small garden,

you can manage it much more carefully. You can enrich the soil, monitor pests, and water plants right when they need it, all of which translates to higher yields. A farmer with many acres of land simply cannot manage the land as intensively.

One example of this is the urban homesteaders at the "Path to Freedom" in Pasadena, California. Their garden boasts a harvest of up to 6,000 pounds of food on just 3,900 square feet of gardens using intensive growing methods. This translates to more than 67,000 pounds per acre! On their small urban lot, they also have chickens, ducks, rabbits, bees, and pygmy goats, as well as graywater and aquaponics (small-scale fish-raising) systems.

GARDEN EVERYWHERE YOU CAN. Small-scale intensive systems can be fit into any area. Patios, porches, balconies, the space between the road and the sidewalk (often referred to as the sidewalk buffer, parking strip, or "hellstrip"), rooftops, and vertical spaces (fences, walls, and buildings) can all host small gardens. There are so many places in our landscape that can become more beautiful and productive. Think of planting berries along a house edge, tea herbs in containers on a patio, insectary plants on the rooftop, fruit trees along the road edge. Even parks can be edible! The Beacon Food Forest in Seattle is a 7-acre plot at the edge of Jefferson Park that has been designed with food-producing plants, and the public is welcome to forage.

DE-PAVE. Urban areas tend to have a lot of pavement — often more than is needed. Pavement takes land out of use for growing food, supporting natural diversity, and capturing stormwater. Excessive pavement also serves as a heat sink, contributing to cities being uncomfortably hot in summer weather. De-paved, the soil can support life in so many ways. Old parking lots, commercial areas, spaces around buildings — these areas can be opened up again, the soil cleaned up and regenerated, and the space made available for people to use, whether for gardening, decorative landscaping, parks, or even just play.

RECONNECT THE WASTE STREAM TO THE SUPPLY STREAM. All the food waste, wood chips, humanure (see Making Use of Human Waste, Safely, page 91), and urine collected from city dwellers could fuel urban farms and gardens, drastically reducing the need for external fertilizers while reducing the burden placed on waste management systems. In some places, neighborhood collection systems could bring food waste to nearby composting operations. The high-value compost would fertilize the urban farms and gardens. City-wide operations are more complex, but they could begin to take advantage of commercial-scale waste streams from businesses and factories. No longer would trash need to be

Using strategically placed curb cuts and stormwater runoff basins helps keep rainwater on-site, where it can infiltrate back into the ground instead of flowing into streams and rivers (often carrying pollutants with it).

A two-acre plot near a derelict building in Detroit serves as a neighborhood farm and a source of inspiration for the community.

trucked to faraway landfills or burned in incineration plants that pollute the air and water.

In addition to polluting, landfills and incineration plants are expensive, and as the population expands, more of these wasteful systems are needed, along with more trucking of waste to distant places. The money saved as cities repair, replace, or upgrade their waste management systems would more than offset the costs to set up these waste recycling systems.

KEEP WATER ON-SITE. Rainwater running off paved surfaces could be collected, stored, and used for irrigation and, if filtered, for human needs, such as drinking and washing. Water runoff from the paved areas in a city heats up and picks up chemicals and sediment along the way to the nearest waterway or water-treatment facility. Channeling that water into on-site vegetated areas keeps it cool, allows it to be used to irrigate crops, and gives it a chance to be filtered before it moves on. Rain gardens do this by collecting and holding the stormwater in a shallow depression in the ground, and they can be planted with a variety of deep-rooted pollinator-supporting plants, berries, and dynamic accumulators (see page 31). In lightly traveled paved areas, permeable pavers (which allow water to flow through them and into the ground) can allow water to infiltrate and keep it on-site.

SUPPORT URBAN ECOLOGY. "Urban ecosystem" may seems like a contradiction in terms, but in fact there's a lot of room in cities for ecology to flourish — along fence edges, in trees, among the woodlands in urban parks and cemeteries, in perennial flower plantings, in rooftop gardens, in vacant lots, and along urban streams. In his book *Wild Urban Plants of the Northeast*, Peter Del Tredici presents a "taxonomy of urban landscapes" and defines the various urban ecology zones as either native-remnant landscapes, managed-constructed landscapes, or ruderal landscapes — those that include abandoned or neglected lands and urban infrastructure that are dominated by volunteer plants that can survive these severe conditions. These "waste" places are actually functioning as ecosystems, providing nutrient cycling, going through succession, catching and filtering stormwater, and providing habitat for pollinators.

As soon as we see the urban landscape in this way, we realize that the world around us, pristine or not, is alive and ever changing, and it can heal from the worst damage imaginable. There are many ways we can support the health of urban ecosystems. Among other things, we can clean up trash along a stream edge; turn an abandoned lot into a meadow with pollinator-supporting plants and fruiting shrubs; plant trees that can handle the urban heat and help detoxify the soil. Every little bit helps.

Eric Toensmeier and Jonathan Bates, along with their families, have created an urban permaculture paradise in their 1/10-acre backyard in Holyoke, Massachusetts.

Before I had even heard of permaculture, I came across the book *The Integral Urban House* (1979). In it the authors, Helga and Bill Olkowski, describe how they transformed their house in Berkeley, California, to become more self-sufficient and consume fewer resources. In the 1970s, with others from the Farallones Institute, a forward-thinking organization founded by Sim Van der Ryn in 1970 (see Resources, page 152), they redesigned and retrofitted the house to grow food, harvest solar energy, recycle wastewater, capture and store rainwater, and raise livestock in an urban setting. The house was a research and teaching center in appropriate technology, energy use and efficiency, urban agriculture, and ecological design.

Their work countered the idea that to reduce their ecological footprint, live more ecologically, and become more self-sufficient, city dwellers would need to move to the country.

The High Line in New York City reclaimed an abandoned elevated rail track, creating urban habitat for wildlife and green space for humans.

GROWING A HEALTHY CITY

There are good things about cities, too! Urban areas provide the opportunity for dense living, allowing us to take up less space, with less sprawl across the landscape. Cities and towns provide centralized markets for collecting and distributing goods, making the transportation of goods more efficient. Public transit systems, such as buses and subways, and a lack of parking space make driving much less desirable, which in turn cuts down on fossil fuel use and air pollution.

Most cities provide for few of their own needs, however, and their "waste" is an underutilized resource. Imagine what it would look like if cities were able to divert this flow of resources back into the urban ecology; it would decrease the need for both food import and waste export.

From Driveway to Food Garden

IN THE CITY, where land is at a premium, sometimes de-paving is the only way to create usable land for growing food. My friend Paige Bridgens, who lives on a small urban property in Northampton, Massachusetts, lacked enough garden space to grow the diversity and quantity of food and herbs she wanted. She also wanted space for raising chickens. Her house had only a small yard, with a full side of the yard taken up by the driveway. The solution was to remove most of the driveway, leaving just enough room for a pull-in space for one car.

Removing asphalt is surprisingly easy. In this case, Paige's driveway paralleled her neighbors' with no separation, so we rented a special saw to cut the asphalt between the two driveways. After cutting, the asphalt came up easily with sledges, iron bars, and rollers to move the large pieces. We took the asphalt to the nearby asphalt manufacturer for reuse, and we dug out the sand underneath and gave it to some neighbors who wanted it. We filled the resulting hole with soil and compost, topped it with rock dust (a waste product from the local quarry) to add minerals, and planted. Within a year, the former driveway was a productive garden filled with a diversity of food, flowers, and pollinator-friendly plants.

It turns out that freeing the soil from its asphalt covering was an easy first step in regenerating the land. Now Paige has more garden space, the water infiltrates the ground instead of flowing into the storm drain, and there's a little more green space in the world.

In this project, we worked with the homeowner to dig up some of the asphalt from her driveway to make space for a garden. The asphalt was recycled and the new open space was filled with compost and rock dust. Within a year, it had become a productive garden.

9

STABILIZE *Our* FOOD SUPPLY

Modern food production is a huge and complex corporate system with many weak points that threaten its stability. The entire system is based on the availability of cheap fossil fuels for producing food and transporting it; our food comes from around the world via ship, truck, and train to get to our store shelves. As such, any disruption in the fuel supply has the potential to cut off the food supply.

Another weak point is that industrial agriculture uses just a handful of plant varieties and animal breeds. This lack of diversity leaves the system open to disruption if a new pest or disease arrives or a livestock breed or crop cannot adapt to environmental changes. In the thousands of years of crop and livestock selection, farmers, horticulturists, and graziers have created a vast agriculture diversity that feeds, clothes, treats ailments, and supports humanity in myriad ways through changing conditions. This genetic trove is threatened, as animal breeds and plant varieties fall out of use.

Unfortunately, once the food is produced, we can't always guarantee that it's getting to everyone who needs it. In many places — especially in economically challenged areas of inner cities — fresh food simply isn't readily available. In these so-called food deserts, residents must travel long distances to reach the nearest grocery store. Finding a way to make sure these people have access to food that is affordable and nutritive is another piece of the food supply puzzle.

To make our food supply more secure and widely available, we must:

MAKE FOOD PRODUCTION PART OF EVERY BUILT LANDSCAPE. Food production shouldn't be limited to farms in Iowa; we can weave agriculture into our residential and commercial landscapes. Imagine the open spaces in our communities filled with a diversity of food, medicine, forage, and fiber plants. The land around residences, commercial buildings, college campuses, hospitals, prisons, and malls can become diverse, beautiful food production zones. On the grounds at Kent Hospital in Rhode Island, for example, you can find healing gardens with relaxing, soothing, and meditative spaces; medicinal gardens growing herbs and plants with medicinal properties; and a community garden that my design firm helped create. Many college campuses are adding food production gardens, which can not only supply the food services department but support curricula and help students connect to the food system.

HARVEST FOOD FROM A VARIETY OF SOURCES. In addition to the usual vegetable gardens, food can come from small plots of grain, backyard livestock, foraged plants and mushrooms, hunting and fishing, aquaculture (see page 66), and nut and fruit trees at the edges or mixed into forests. Not only does this diversity allow for a wide array of crop choices, but it also builds resilience into the food supply. If one crop fails because of pests or adverse weather conditions, other crops can fill in the gap. Learning to eat from a diversity of sources creates a connection to the places and people that provide food for us.

Even our ornamental landscapes can produce food! Most people don't realize, for example, that kousa dogwood (*Cornus kousa*) trees produce delicious fruits in the fall.

EAT SEASONALLY AND LOCALLY. Varying crops also means that fresh food can be harvested throughout the year. In my family's forest garden, early strawberries, Nanking cherries, and early perennial vegetables such as sorrel and Good King Henry are followed by raspberries, blackberries, and garden greens, which are followed by the summer pulse of currants, gooseberries, and annual vegetables. Then come the late-season crops: apples, pears, persimmons, hickory nuts, walnuts, and squash. Each season has its bounty, and when you learn to eat what's in season in your region, you don't need to depend on global transportation systems to get food to you. In this way, eating seasonally leads to eating locally, which helps strengthen the local economy.

RAISE HERITAGE ANIMAL BREEDS AND GROW HEIRLOOM PLANT VARIETIES. Heirloom varieties and rare breeds represent the work of generations of farmers and growers who selected traits to help these species through hard times and to adapt to various conditions. Scottish Highland cattle, for example, are a rugged breed that has adapted to survive cold winters on poor-quality feed. Unfortunately, our agricultural biodiversity is at risk, as many animal breeds and plant varieties are being lost and forgotten in favor of popular commercial varieties. Often these commercial varieties sacrifice flavor, hardiness, and other positive qualities for the ease of shipping and long shelf life. We can keep our food supply diverse and resilient by continuing to plant heirloom varieties and by purchasing products from farmers who do the same.

Diversifying our food supply, by growing heirloom crops and raising heritage breed animals like the Scottish Highland cattle, helps ensure its resilience.

Sirius Community Center, Shutesbury, Massachusetts

BIOSHELTER: A system of food production in a passive solar greenhouse that integrates growing beds, animals, aquaponics, use of vertical space, heat storage, and energy efficiency to create a designed and managed indoor ecosystem.

PROVIDE FARMERS ACCESS TO AFFORDABLE, QUALITY LAND. Farmers need land. And finding affordable land with fertile soil and available water that is close to customers (with housing on-site or nearby) is a big challenge. Farming is hard work even without the difficulty of getting and holding on to land. With the growing interest in small-scale farming, securing land near a customer base is difficult and financially stressful (in an occupation that is already financially marginal). There are many creative ways to access land. Some communities are banding together to protect agricultural land in danger of development and then leasing or selling it to farmers. Increasingly, land trusts are making it part of their mission to protect agricultural land and make it available and affordable. Many aging farmers are willing to find creative ways to pass their land on to younger farmers. In addition to better access to land, farmers also need access to seed, organic fertilizers, tools, and technical support in organic and ecological practices.

FISH IN THE BACKYARD

Aquaponics (the name is a contraction of the terms "aquaculture" and "hydroponics") is a method of raising fish and growing plants in a closed-loop, low-input system in which the nutrient-rich water from the fish can irrigate and feed crop plants. The plants take up the nutrients, utilize them, and in the process filter and cleanse the water. These small-scale, intensive production systems are often part of a "bioshelter," a complex and integrated greenhouse system with production beds, aquaponics, livestock (often chickens and/or worms), water collection, and vertical growing arrangements, all meshed together. These integrated structures help extend the growing season and make urban spaces more productive, even in winter. Growing plants with fish, crustaceans, and other animals in a tightly cycling water system mimics the ecosystems of wetlands, estuaries, and ponds and produces food, cleans water, and provides a beautiful, warm space to be during the long winter.

At Alchemy Farm, formerly New Alchemy Institute, Hilde Maingay cares for the plants in the bioshelter.

A bioshelter creates the right climate for indoor growing and for raising fish in small ponds.

Create PRODUCTIVE LANDSCAPES

Imagine that wherever you look across the landscape, food is growing. The town center hosts a community garden. The local elementary school has fruit and nut trees in the fields surrounding them. Businesses have replaced ornamental landscapes with berries and herbs.

Lawns are dotted with vegetable gardens, chickens, fruit trees, and lively edible landscaping. This scene is becoming more real in many communities across the country. Spaces that were once mowed or otherwise underutilized are producing food. We really can grow food just about anywhere!

One focus of permaculture is to create yields on the land we live on, in our own communities; this means we're able to take care of our needs close to home. Many benefits flow from this ideal, not the least of which is healthy, fresh food. When we produce and sell food locally, money cycles through the community, providing jobs and fostering a robust local economy. In fact, local production of foods of all types (vegetables, fruit, dairy, meat, grain), as well as medicinal herbs, tea, fuel (such as firewood),

building materials, and even clothing, shoes, baskets, and soap, is a powerful economic engine to revitalize local economies. In Ben Hewitt's 2010 book, *The Town That Food Saved*, he relates the story of Hardwick, Vermont, and how the town's embrace of local-food-based agriculture and marketing sparked its transformation into a thriving community with a vibrant local food system.

Productive landscapes not only benefit those of us who live there but also take the pressure off the distant agricultural lands that currently grow food for the global market. For instance, land in the Central Valley of California, a major breadbasket of North America, produces huge amounts of food for tables across America and around the world. But the cost of this production is devastation of

Right: Asian pears suffer from less pest pressure than European pears do, so they're easier to grow and more attractive in the landscape, while being just as productive.

Below: Shady areas that aren't ideal for cultivating plants can be used for growing mushrooms.

Chickens help keep pests at bay and leave behind manure to fertilize plants.

this incredibly rich valley ecosystem, once home to myriad species and historically a central stopover and wintering ground for huge flocks of migratory waterfowl. As the land has been drained and leveled for agricultural use, its diversity, regenerative capacity, and fertility has suffered. These days, almond and olive orchards, row crops of sugar beets, and rice fields dominate the landscape.

As it turns out, where I live in the northeastern United States is actually a great place to grow rice.

The more rice we grow here, the less rice needs to be grown in California on important wetland edge habitat. Growing rice in the Northeast can also mean that we're able to use low-lying agricultural lands as they are, instead of draining them for crops that need drier conditions. Maintaining wetland areas and restoring those that were drained for use in agriculture or development both rebuilds local agricultural systems and benefits local ecosystems.

In addition to producing food, landscapes can also grow a bounty of materials for other purposes, including coppiced willow for wattle fencing, locust poles for fenceposts and pole-built structures, and wood for biochar.

So-called *terra preta* ("black soil" in Portuguese), found in the rainforests of Brazil, is extraordinarily rich soil – a result of charcoal deposits incorporated into the earth by indigenous peoples. The discovery of terra preta has lead to research into the benefits of biochar for soil fertility and for sequestering carbon in the soil.

HELP REVERSE
Climate Change

By now, it's clear that the earth is heating up. The specific long-term ramifications are uncertain and dependent upon uncountable factors, but the future does not look good for humankind. Much of the warming can be linked to a buildup of the vast pool of so-called greenhouse

gases in the atmosphere. These gases allow sunlight to penetrate the earth's atmosphere and reach the surface of the planet. Some of that light is radiated back, and the greenhouse gases absorb it, thus trapping it in the atmosphere. Some greenhouse gases are naturally occurring, such as carbon dioxide, nitrous oxide, and methane, while others are man-made, such as fluorocarbons.

Heat trapped by this blanket of greenhouse gases, primarily carbon (in its gaseous states), is threatening to change the fundamental climatic conditions on the planet. Atmospheric concentrations of carbon and other gases have been rising sharply since the time of the industrial revolution, and as those concentrations build, the speed of climate change will increase. According to data from the U.S. National Oceanic and Atmospheric Administration, atmospheric levels of CO_2 have increased from 280 ppm (parts per million) prior to 1850 to more than 402 ppm as of January 2016. We've gone over 400 ppm

for the first time in millions of years. Some scientists believe that we need to quickly reduce atmospheric carbon levels to 350 ppm to stop severe climate change.

We need to act quickly and decisively. Of course, the most important action we can take is to emit less carbon into the atmosphere. This means changing our actions and habits: driving less, buying locally produced products, engineering deep energy retrofits for buildings. Reducing industrial emissions for energy production is critical, too.

Although reducing carbon emissions receives much of the public focus when climate change is discussed, another way to reduce the level of carbon in the atmosphere is to put it back into the ground (a process referred to as "carbon sequestration"). Soil holds more carbon than the atmosphere and all plant life combined; in fact, 80 percent of all terrestrial carbon is in the soil. Only the ocean has more carbon storage.

Carbon is held in the soil mainly as a component of organic matter — plant and animal material in the soil in some stage of decomposition. Rich, fertile soils have lots of organic matter and, thus, hold lots of carbon. Poor soils and degraded soils don't have much organic matter and thus do not serve to sequester carbon. The bad news is that the clearing of forested land, reductions in soil organic matter, and decomposition of aboveground organic matter decrease soil carbon, releasing it into the atmosphere. Soils from degraded farmland, urban river and stream channels, and suburban lawns fertilized with chemicals have only a fraction of the fertility and carbon they once held. It's estimated that the world's cultivated soils have lost 50 to 70 percent of their carbon.

The good news is that our food production and forest management systems can be converted to sequester carbon just by altering the way we farm and the way we manage land. Increasing soil organic matter by just 1 percent on the world's estimated 14 billion acres of agricultural land would sequester 256 billion tons of carbon — three-quarters of the CO_2 we need to sequester to get back to 350 ppm.

To sequester carbon, the permaculture "carbon farming" toolbox focuses on integrated agriculture and land management strategies:

- Enriching soil carbon through better agricultural practices, such as reducing (or eliminating) tillage and using cover crops

- Farming with trees and other perennials

- Climate-friendly livestock production; management-intensive grazing, not overgrazing; and using silvopasture (systems that combine trees and pasture)

- Protection and restoration of natural habitat

- Restoration of eroded and degraded lands

In an integrated system — one that uses several carbon sequestration practices — farming and land management can sequester more carbon than they emit. Growing deep-rooted perennial plants is one of the primary strategies. Perennial root systems produce nine times more roots than annuals, building carbon in the soil as the root systems become established and shed from the plant. So why not have our carbon sequestered and get nuts and fruit, too?

Because these practices have multiple benefits beyond carbon sequestration — such as growing food, restoring healthy ecosystems, and supporting wildlife — it makes sense to start by restoring degraded and damaged lands. There are vast tracts of denuded lands across our landscape, in rural, urban, and suburban areas. This is where we need to start. Some people may advocate converting forested land to agriculture for carbon sequestration, but this is immediately a net loss. A better plan is to start with those lands that have lost their carbon storage capacity, and restore their role in the carbon cycle while at the same time fostering their productive use to support humans, wildlife, and nature as a whole.

Increasing soil organic matter by just 1 percent on the world's estimated 14 billion acres of agricultural land would sequester 256 billion tons of carbon — three-quarters of the CO_2 we need to sequester to get back to 350 ppm.

~~~~~~~~~~

12

HELP YOU
Become a Nicer Person

No whole system can ignore the personal and interpersonal aspects of our lives. The term "permaculture" — although it began as a contraction of "permanent" and "agriculture" and was originally focused on land and ecosystems — has come to encompass the whole of society and a broader definition of what it takes to create a "permanent culture."

Our life extends well beyond what we *do* (whether we can grow vegetables, build a greenhouse, or plumb a graywater system). We have an inner, personal life that directly affects how we move through the world, how we navigate our relationships, and the roles we have in our community. Our internal motivations, our personal and family history, and our emotional and mental clarity can work for us or can sabotage us. The work we do in the world must include our own inner work — the work to free ourselves to be full and whole members of our families and communities. Whether or not we acknowledge it, our inner life affects the world around us and the way we perceive the world.

This became especially clear to me in 1996 when my wife, Kemper, and I moved to Lost Valley Educational Center, an intentional permaculture-oriented community in the foothills of the Cascade Mountains of Oregon. We were very excited about living on the land, growing food, and doing all the great work that goes with it. I was focused on the hard skills and the land work to be done, but within a short time the complexity and challenges of living closely with other people began to surface; personal issues came into play, hidden feelings surfaced, arguments arose. I quickly saw that the work of living in a community with others was going to be a big challenge. Clear communication was utterly necessary, as was identifying and addressing the internal processes that were blocking me from being my full and present self, living closely with others.

At Lost Valley, we had a weekly business meeting, which focused on the businesses we

Whether or not we acknowledge it, our inner life affects the world around us and the way we perceive the world.

ran together, and a weekly community meeting, which focused on interpersonal and community issues. It became apparent during the many hours in meetings, circles, and side conversations with my fellow communitarians that living together was the most difficult aspect of life in a community. Hosting educational programs, growing food, raising kids, even running businesses together was relatively easy compared to the work of staying present and clear in relationships. It also became apparent that once the inner work was attended to (an ongoing process), the outer work of life could proceed easily and with grace.

If we don't attend to the interpersonal issues, then myriad troubling, difficult, sad, and intrusive behaviors arise and interfere with our relationships. These can keep us from becoming our best self, and prevent us from recognizing our own gifts and being able to share them with our community. It takes time and effort to build the relationships and trust required to regenerate community. Over time, as the bonds develop and strengthen, people begin to feel comfortable leaning on each other in times of need; sharing surplus of time, food, or energy for work; supporting each other as children grow to adults or elders pass away — through all the joys and sorrows of life.

PRACTICING OBSERVATION FOR MINDFULNESS

Focused observation, in nature and within ourselves, is a core practice for creating awareness in our everyday lives. You can observe the land for animal tracks or water movement, and also internally "track" your thoughts, reactions (or nonreactions), and emotional states. Open observation is a practiced skill of looking without judgment or labeling. When we observe openly, patterns emerge that we would miss if we quickly scan and identify or label whatever we see. In the practice of nature awareness, this wide-angle vision, sometimes called "owl eyes," helps us stay open to all input without putting a name to it. When we label and box things and project our own perceptions of "good" or "bad" or even just attach our own definitions to them, we miss the subtleties and possibly some very important information.

The ponds at Ben Falk's Whole System Design site hold the water high upslope to be available for irrigation and groundwater recharge. The pond by the sauna (above) also moderates the microclimate for the gardens nearby.

Help You Become
A BETTER DESIGNER
(of Landscapes and of Life)

Design is a thoughtful and intentional process, and the beauty and life in well-designed places are easy to recognize. Christopher Alexander, in his books *The Timeless Way of Building* and *A Pattern Language*, calls this the "quality that has no name." It's a sense of aliveness in communities and homes.

These are the places we gravitate toward: a community center with grass and trees; a shop-lined street with broad sidewalks and tables for eating a meal and chatting; a rolling landscape with farm fields, houses, and a small town center. Connectedness of place comes when buildings are built to scale with the people living there, the spaces are welcoming and easy to move through, and water flows freely and trees cast shade onto streets and gathering spaces. People gather in these well-designed and welcoming spaces.

Unfortunately, all around us we can see the results of senseless and destructive design that focuses on quickly creating things that are cheap and disposable and not at all welcoming. Our homes, businesses, and infrastructure are rarely built to last the test of time and to be healthy and safe to live in. Our communities sprawl across the land, encourage wasteful energy use, consume vast resources, and destroy ecosystems.

RECLAIMING OUR ROLES AS DESIGNERS

For the most part, we accept our surroundings as they are and feel that there's little we can do to change them. We relegate the role of "designer" to other people. We seem to have forgotten that we are part of the process — we are all designers with an inherent design sensibility. We need to reclaim our right to create the homes, landscapes, and communities we live in. In doing so, we can make the built world around us more beautiful, full of life, and reflective of our personal values.

EVERYONE IS A DESIGNER. Acknowledging our own ability to design is just the first step. Then we must learn to listen and to see the world around us as a patterned whole. Every piece of land has a story to tell. Our work is to listen for that story, so that we can identify the potential of the place and determine whether it's right for the activities we'd like to carry out there. Listening to the land will also help us identify limitations or problems that need to be addressed. Keen observation will shape a better understanding of the place as a layered whole.

OBSERVE. Try this: Step back, let your eyes go wide, and take in everything you see — including the slope of the land, water, paths and access ways, structures, vegetation, animals, sun patterns, and airflows. As you tune your senses and cultivate your powers of observation, the land will begin to reveal its patterning: the subtle slope changes, wind flows in connection to the land, sunlight patterns at different times of day and year, the way certain plants grow better in a particular place, and how they clump together in one spot or are spread more evenly in others. Notice where areas of high activity are, where the soil is deep and rich and where it is exposed, and how that affects the plants growing there. You might notice that a low area on the land has slightly different plants; a hole dug there might reveal groundwater near the surface, or evidence of rising and falling groundwater.

NOTICE LEVERAGE POINTS. By seeing the land as a patterned whole with all its complex connections, we start to see *leverage points* — places where we can make small changes and have big effects. These are good places to start. A leverage point might be capturing and storing water in a swale — simply a low place you create in the landscape to hold water and allow it to slowly infiltrate the soil rather than run off and create erosion. A swale not only decreases soil erosion, it also creates a minienvironment that fosters life. In some cases, this minienvironment can be the keystone to bringing life back to desertified regions.

This is an example of how a permaculture designer would identify patterns in the landscape. But just think of what it would look like if you applied this kind of observation and pattern thinking — what designer Christopher Alexander called (and titled his book) "the timeless way of building"— to every aspect of human society, starting with your own life. What are the leverage points that would make dramatic, positive changes in your life and your community?

Reclaiming our role as the designers of our own lives allows us to reweave the tattered connections in our homes, land, and communities. It allows our children to grow up with the experience of aliveness of place and caring about the land and the communities they live in. And as we slowly repair our communities and culture, we build momentum against the tide of unraveling.

DESIGNING WITH NATURAL PATTERNS

Patterns are around us everywhere. The branching pattern of the tree above- and belowground efficiently collects nutrients and water from the soil, sends them up through the trunk, and distributes them to feed the leaves and fruit. Branching patterns are designed to collect or distribute from/to a wide area. Mimic this pattern in laying out a garden path system to send compost from the compost pile out into various garden beds or to direct traffic through a city. A branched path pattern will efficiently distribute the material throughout the garden.

A more orderly grid pattern is useful for easy navigation and getting readily from any point to any other point. Grids are often used, even at a small scale, for production beds and places where crop rotations can be planned with quick area calculations. Think of the grid layout in a community garden or a city like Los Angeles. This pattern allows lots of access for the many people getting to and from all the various plots to the compost pile or a wash station or an arrival space.

MAKE SMART, LONG-LASTING CHOICES

Sometimes I visit a property for a consultation and just look around, wondering where to start to make sense of it all. Given the complexity of the land, one powerful tool to tease apart the layers is the "scale of permanence" – a site assessment tool that orders a site's components according to how difficult it is to change them. Then, when you're considering how to work with the site, you should consider first those aspects that are most difficult to change. The following list of elements, from most permanent to easiest to change, was compiled by David Jacke and Eric Toensmeier in their books, *Edible Forest Gardens (Volumes 1 and 2)*, and is based on original work by P. A. Yeomans:

- Climate
- Landform
- Water
- Legal issues
- Access and circulation
- Vegetation and wildlife
- Microclimate
- Buildings and infrastructure
- Zones of use
- Soil (fertility and management)
- Aesthetics

For instance, if you're planning to make changes to the landscape, it's best to begin with the landform and waterways. Add swales, a pond, or terraces – very permanent aspects of the land – before siting a building, for instance. This way, the building is not an obstruction to any grading or slope changes that need to be done and can be placed away from waterways. At one farm I've worked with, someone long ago made the classic error of placing a garage/shop building directly in the path of a water channel. This will be very difficult to deal with without major grading changes and is a problem that, without massive effort, will persist into the future. By simply observing the landform and the movement of water on the site, and then placing the building 50 feet one way or the other, the builder could have avoided such a problem.

It's also advantageous to identify the access and circulation patterns before siting a building. If the building is sited in a spot away from the most efficient access road (working with the slopes and water flow), that building will be hard to get to, and the road to it will be more difficult and expensive to maintain. Many people, unfortunately, site a home based on the view from a location – even if it means putting the home at the highest, windiest, hardest-to-reach point on the land. Then the driveway is designed to get there – even if it means going through sensitive woodlands or wetlands or up a steep slope.

Simple design choices – like siting a house partway up a hill instead of in the valley, where it could be flooded from water flowing down the hill – can make a huge difference in a structure's longevity and how it functions in the landscape.

We made small, incremental changes to our property over time to increase the food-producing potential of the land, to improve the energy efficiency, and to open up more of the south side of the house to the winter sun.

Before

After

(14)

BUILD
Smarter Homes

Home is where the hearth is. Much of our lives is spent in our homes, which can be a fulfilling experience or an uncomfortable one. Modern homes are built with little attention to passive solar potential, energy efficiency, nontoxic materials, or land connections. Permaculture is all about making the connections through design. The home is a great place to make this happen.

Let's explore this idea by taking a look at the work my own family did with our home, Hickory Gardens, in Leverett, Massachusetts. When we moved there more than a decade ago, there were many things we loved (and still love) about the land: a very short driveway, which meant little maintenance or snow to remove; a generally east–west orientation (good for passive solar), and close proximity to neighbors and the community co-op/grocery store. And because the house is naturally earth-bermed, with a ground-level entrance, it's protected from the prevailing winter winds, and heating and cooling are buffered by the temperature of the ground.

But we also faced a common situation — the house had been built with little consideration of energy use. On top of other problems like questionable wiring and plumbing that was held together with duct tape and bungee cords, there was very little insulation. The windows were drafty, and most of them faced the road to the north. It was a neighborly arrangement, but because our winter storms come from the north, the house was very exposed to cold weather. We didn't have the resources to build a new home somewhere else or to take down our old house and start over. The only choice we had was to slowly retrofit the house and make it more energy efficient while we lived in it.

KEEP OUT THE COLD

Our first priority was the basement. The walls and floor were uninsulated concrete, leading to condensation and mold. To begin, we added insulation. We also lowered the ground level outside a bit, just enough to allow for short basement windows on the south side, giving more natural light and airflow. We used the open walls to rewire, replumb, and put in forced hot-water heat — a very efficient system that produces even, comfortable heat and can eventually be connected to a solar thermal system.

Early on, we enclosed part of the front porch for a mudroom. Instead of having cold air blow through the front door in the winter every time we come in, we enter into the mudroom, close the outside door, and remove coats, hats, gloves, and boots. The mudroom is insulated but not heated and acts as a buffer between the outside and inside. The cats also come through the mudroom, using a series of doors (but smaller). They can wipe their feet and drop off mice and voles before coming in!

UPGRADE OVER TIME

Over time, we've worked our way around the house, insulating the walls and replacing old windows. A few years ago, we were ready for a big upgrade to the south side of the house. A dilapidated, uninsulated south-facing porch was blocking light and leaking cold air into the house. Water pipes in the outside wall would freeze. We rebuilt the new space into a large family room and took the opportunity to adjust the house layout and fix some other key problems in the process. The bathroom, in the center of the house, moved to the northeast corner, and the funky plumbing was replaced. The new bathroom has the clothes washer for easy access (we don't have a dryer, preferring an outdoor clothesline in summer or a drying rack by the woodstove in winter). Instead of replacing the flush toilet, we installed a composting toilet so our humanure can be recycled on the land. (See page 91 for more on humanure.) Graywater goes into the septic system for now, but a future project will be to make use of that

Removing the uninsulated, south-facing enclosed porch and replacing it with an insulated, window-filled space allowed sunlight to enter our living space and made the whole house warmer.

Increasing the amount and quality of insulation in a house is one of the best first steps a homeowner can make.

resource. Leachate from the composting toilet goes into the septic system by law, but should the right time come, we can collect that nutrient-rich water and use it to build fertility on the land.

The walls of the addition are 10 inches thick and have blown cellulose (recycled jeans) insulation. Insulation in the ceiling is 12 inches thick, and an 18-inch-thick blanket of insulation in the attic space reduces heat loss there. With the bathroom removed from the center of the house, we installed the woodstove there, and it now serves as the central heat source. Heating with wood from our land allows us some control over our heat supply. The increase in insulation and closing off the north walls has reduced our winter heating requirements to about two to three cords of wood each winter.

Summer cooling is easy in a well-insulated house. We open the windows at night to let in the cool air. In the morning, as the outside temperature is warming up, we close the windows and the house stays cool all day. A ceiling fan gets the air moving on the hottest days. Wide house eaves and trees to the west block midday and afternoon sun and reduce the afternoon heat. And a dip in a nearby stream or pond on the hottest days helps, too!

SECURE THE WATER SUPPLY

To add redundancy to our water supply, we built a 3,100-gallon cistern into the basement of the new space, simply by pouring an extra wall at the end of the foundation. The cistern stores rainwater collected off the roof. The water is gravity-fed to the backyard gardens or can be used in a future home aquaponics system; it's also available for emergency use when we lose power, which occurs occasionally in winter storms, sometimes for several days. This water source provides backup for our well and requires no pumping or power to operate.

OTHER STRATEGIES

In addition to everything else, we've replaced drafty windows on the north side of the house, improved the insulation in the existing walls, and replaced appliances with energy-efficient models to reduce electric use. We've also installed low-flow water faucets and showerheads whenever possible.

All this work had to be done before we could consider investing in solar electric or solar hot water systems. Those are great technologies, but for us, the first steps were to lower our energy needs.

As with any house, there are aspects to an energy retrofit that we couldn't do either because the home and land were not suitable or we couldn't afford it. Some of the other strategies you might consider:

WINDBREAK TO BLOCK PREVAILING WINTER WINDS. Large energy savings can be gained with a windbreak to reduce cold winter winds. The cold air pulls the warmth from a house, particularly if it's not well insulated. In the right situations, a windbreak with several rows of evergreen trees can reduce energy use by 25 to 30 percent and even up to 50 percent.

ATTACHED GREENHOUSE. An attached greenhouse is sometimes a good way to passively heat a structure. It is important that the greenhouse be built outside the thermal envelope of the structure, to allow control of the air movement between the greenhouse and home. Humidified air should not be allowed inside the home in a high-humidity summer environment. This can lead to mold and mildew spores in the home. But a well-designed greenhouse can make it possible to grow plants through the year, add light and warm air to the home, and offer a getaway from the cold. It can also potentially hold space for aquaculture, worms, chickens, and more.

Someday, we may be able to attach a greenhouse like this one to the house, to extend the growing season and help reduce our heating needs in the winter.

MAKING USE OF HUMAN WASTE, SAFELY

People have been safely turning human waste – often referred to as "humanure" – into valuable compost for thousands of years. And not just their own! In his 1911 book, *Farmers of Forty Centuries*, agronomist F. H. King documented the farming practices of peasants in Japan, Korea, and China; he witnessed farmers building decorative outhouses meant to tempt travelers into utilizing them so they would have more fertility to add to their fields.

But in our current day and age, using human waste for any purpose is a cultural taboo. Instead, we use the flush toilet, which takes the waste to a waste-treatment plant or, if you live in a rural place, to a septic field in your backyard. Either way, instead of being utilized, this resource becomes a source of pollution.

The energy cost of pumping wastewater to a treatment plant and treating it is immense, and the chemicals used in treatment are then spilled into the nearest body of water, along with the partially treated waste. Waterways overloaded with the additional nutrients from the waste spawn algae, which leads to oxygen loss and dead zones in rivers, lakes, and oceans. Septic systems are no better; they can leach nutrients into groundwater and have the potential for polluting nearby waterways. On top of all that, the water that fills toilets is fresh, drinkable water that we turn into wastewater.

There is an alternative, though. There are safe ways to compost humanure, either with simple home composting toilets or with more complex neighborhood or town-wide systems.

We use a composting toilet in my home to effectively convert our family humanure into compost over the course of a year or two. State laws are changing and making it easier to do this. We use the compost in our orchard or forest gardens to enrich the soil and cycle the nutrients back into the food we eat.

An example of advanced thinking on recycling nutrients on and off farm is in Veracruz, Mexico. At the Las Cañadas community, cooperative members who don't live on the land must install a compost toilet and bring back a certain amount of composted humanure when they come to pick up their harvest share. In this way, they replenish what is lost by farm production being sent off farm. This is another method to close the loop of nutrients and organic matter and make use of human waste instead of treating it as a problem.

15

CREATE GARDENS

That Provide for

THEMSELVES

Too often, we see each part of the natural landscape as separate and individual. In reality, all parts of nature are interconnected.

A bird eats a cherry off a tree. It flies across the field and deposits the seed where it lands. The seed sprouts into a cherry seedling, which roots deep into the soil, breaking up a hardpan layer so that other plants can take root. As the cherry tree grows, it creates shade beneath its branches, allowing several herbaceous understory plants to get started, and these begin to improve the soil conditions and support insects that pollinate this and other plants. The cherry tree sets fruit, and a bird comes to eat them, and the story continues.

Likewise, in the landscapes we create for ourselves, we tend to treat plants as individuals that offer a specific yield to us — fruit, vegetables, firewood, fiber, flowers, shade. We don't take into consideration the benefits they offer to each other, or how, by treating them as groups rather than as individuals, we might be able to reduce maintenance, increase our yield, and create a well-functioning ecosystem that can sustain itself.

In order to create a productive, self-sustaining landscape, any plant you include should function in several ways. For example, a groundcover like strawberry not only produces delicious fruit early in the season, but it also shades the soil for other plants nearby, reducing evaporation and suppressing weed growth, and it provides nectar and pollen for insects. Other plants might yield an edible harvest while mitigating soil toxins, or provide erosion control while serving as habitat for native pollinators.

Planting in "guilds" — groupings of plants that are not just multifunctioning but also actively beneficial to each other — goes one step further and leads to reduced work to manage the plants, like weeding and watering, and fewer inputs of energy, nutrients, and water.

Once you define the goals of your polycultures — produce food, reduce weed pressure, attract pollinators, filter dust from a neighboring farm, and so on — you can select species that will work for the location and goals you have defined. Plant ideas can come from books on perennial, multifunctional plants, websites, nurseries (see Resources, page 152, for suggestions), and from visiting local projects doing similar work. Sometimes getting the "keystone" or central plant, such as a fruit tree or shrub, identified can help you get started. Find plants that complement and support that central plant. These "mutual support guilds" create a web of interconnections, a designed ecology that produces abundantly and reduces our management and maintenance needs.

Productive, self-sustaining landscapes should include plants that function in several ways.

LEARNING TO PLANT IN GUILDS

When people plant in guilds, they often weave in nitrogen fixers from the legume family (Fabaceae), pollinator-supporting plants – often from the aster family (Asteraceae) or carrot family (Apiaceae) – and groundcovers to reduce weed competition and hold moisture in the soil.

Each of the guilds below has a shrub, tree, or perennial that is the focus of the grouping. The other plants in the guild are designed to support the central element and provide benefits and yields of their own.

Elderberry Guild

CENTRAL ELEMENT: Elderberry, a medium-size, multistemmed shrub with fruit (delicious cooked) and medicinal flowers. Supports the elderberry longhorn beetle, which is a rare species in some places.

OTHER GUILD MEMBERS: Comfrey, with several species ranging from 18' to 5' high, is a strong groundcover with deep roots that bring up nutrients and feed the soils and other plants. Can be cut several times a season for mulch material.

Green-and-gold (*Chrysogonum virginianum*) is a part- to full-shade groundcover that will grow under the elderberry and reduce weed competition. Supports nectary species (those that provide nectar for pollinators).

Peach Guild

CENTRAL ELEMENT: Peach tree (or another fruit tree); for fruit, the best size is a dwarf tree. For shade, you can select a taller, spreading cultivar.

OTHER GUILD MEMBERS: False indigo bush (*Baptisia* spp.) is a medium-size leguminous (nitrogen-fixing) shrub. It builds soil fertility and can be coppiced for woody biomass and grown around the edges of the central tree.

Currant and gooseberry shrubs grow 2' to 6' and produce prolific nutrient-rich fruit. Black currants are an especially good source of minerals and vitamins, particularly vitamin C. Partial shade tolerant.

Daylily is a spreading groundcover with beautiful, edible flowers. Can take some shade.

Asparagus Guild

CENTRAL ELEMENT: Asparagus is a medium-height (3' to 5'), deep-rooting perennial vegetable.

OTHER GUILD MEMBERS: Strawberry is a low, spreading groundcover ideal for covering soil under and around asparagus, reducing moisture loss and weed pressure. Delicious fruit.

Illinois bundleflower (*Desmanthus illinoensis*) is a medium-height (to 4') leguminous (nitrogen-fixing) herbaceous perennial. Improves soil conditions, and has a high-protein seed.

Gooseberry is tolerant of partial shade and makes a good understory plant in a peach guild.

Strawberry and asparagus work well in a guild together; strawberry provides a low, noncompetitive groundcover for the asparagus, which helps retain soil moisture and reduce weeds.

PLANT GUILDS

Nanking Cherry Guild

CENTRAL ELEMENT: Nanking cherry (*Prunus tomentosa*), a medium-height, multistemmed shrub, offers beautiful early flowers followed by tart cherries.

OTHER GUILD MEMBERS: Alpine strawberry, a low, spreading groundcover with small, tasty fruit, reduces weed competition and increases water infiltration.

Skirret is an herbaceous perennial, 2' to 3" tall, in the carrot family that supports pollinators and has an edible root.

Sweet cicely, an herbaceous perennial in the carrot family, supports specialist and generalist pollinators. Edible leaves, roots, and seedpods taste of aniseed; a medicinal plant, too.

Yarrow, an herbaceous perennial in the aster family that supports specialist and generalist pollinators; an important medicinal herb.

Pear Guild

CENTRAL ELEMENT: Pear tree; for fruit, the best size is a dwarf tree; for shade, you can select a larger variety.

OTHER GUILD MEMBERS: Oregano functions as a low-spreading groundcover and a culinary herb; prolific flowers support pollinator insect diversity.

Chives are small bunching green onions for cooking with flowers that support pollinator activity.

White clover is a low groundcover that fixes nitrogen in the soil (improving the fertility), holds in moisture, and has flowers that support pollinators.

Echinacea or purple coneflower is a bunching flower providing nectar for butterflies and other pollinators; a beautiful plant for the garden.

Raspberry Guild

CENTRAL ELEMENT: Raspberry comes in cultivated and wild species that spread by suckers; its delicious fruit is highly nutritive.

OTHER GUILD MEMBERS: Autumn olive (*Elaeagnus umbellata*) is a non-leguminous nitrogen-fixing shrub that grows to 10' tall; offers good coppice. Its fruit is a small berry high in vitamins and containing very high levels of lycopene, a carotenoid antioxidant. For coppice, plant one shrub every 15 feet, and cut it yearly to 5' to 7' tall.

Asian pear guild

SNAIL PROBLEM OR DUCK DEFICIENCY?

It has been said that waste is just an underutilized resource. As Bill Mollison, cofounder of permaculture said, "You don't have a snail problem, you have a duck deficiency!" Connect the right problem (yield) to the right place and it becomes a resource.

On a farm, the fields yield forage and need compost or manure for fertility. Conversely, livestock need forage and yield manure. Together, fields and livestock form a symbiotic pairing. When modern agriculture separated animals from crops, we suddenly had manure polluting groundwater and rivers and then fields requiring chemical fertilizers. Reconnecting these two gives us the functional interconnection where the yield of one – forage – supplies the needs of the other – livestock – and vice versa. The more we make and support these interconnections, the more we weave a resilient ecosystem.

As the yields of one element are connected with the needs of another, ecological integrity – the soundness and wholeness of an ecosystem, coupled with its ability to handle change and disruption – develops. By mimicking the connections of an ecosystem, we begin to establish relationships and interconnections that lead to stability in our systems and reduced work for us. In our landscapes, gardens, and fields, when we select plants to be mutually supporting, they can be considered a guild – our ultimate aim in a designed ecosystem. This is companion planting taken to a new level.

ECOLOGICAL INTEGRITY: An ecosystem has ecological integrity when the structure, composition, and function are not impaired by stresses from human activity; ecological processes are intact and self-sustaining; it is self-renewing; it is resilient and responds positively to change.

Connecting needs and yields is an important concept in permaculture thinking. You could consider slugs to be a pest problem, but if instead you think of them as duck food, you can meet two needs – the need to reduce the population of slugs in a field and the need to provide duck food – at once.

STOP EROSION & MAKE WATER CLEANER

In most towns, cities, and developments, water is moved off the land as quickly as possible. Although it's critical to control flooding, when water is quickly funneled into rivers and streams without first infiltrating the ground, it can cause rivers to rise quickly and flood downstream areas, decrease groundwater levels, and begin a cycle of stream channel erosion.

To make sure we have enough clean drinking and irrigation water, and to keep the force of water from eroding the landscape, permaculture has a simple mantra: "Slow it, spread it, sink it" — that is, catch the water, hold it on-site, and get the water back into the ground. In the process of catching and distributing the water, we hold it in catchments like ponds, tanks, cisterns, and swales and use it (for growing food, irrigation, drinking, and so on) before it infiltrates the ground. There are a few ways to do this.

STORE WATER IN SOIL. Soil is by far the largest reservoir of water on the landscape, and given the choices of water storage, increasing the soil's water-holding capacity is the cheapest option. Soil that is deep, friable (open and loose, not compacted), and high in organic matter (full of broken-down and decomposed plants and animals) will hold large amounts of water. We can increase water storage in the ground in a number of ways: by building up the level of organic matter in the soil, which improves its ability to hold water; by increasing the amount of vegetation on the land; and by creating earthworks, like swales and basins, that can increase infiltration and recharge the shallow groundwater.

STORE WATER IN VEGETATION. Thick vegetation can hold large quantities of water in several ways. When rain falls into an area with lots of plants, some of the water is held by the leaves and branches, and some is held in the low-growing vegetation close to the ground. (Think about walking out across a field or lawn after a storm. Even when the air has dried, there is quite a bit of moisture in the grass you're walking through.) Some of this water is slowly released into the soil; some evaporates back into the air. Water is also stored in the living biomass of the plants (which are 90 percent water by weight). In thick vegetation, this amounts to a large reservoir of water.

STORE WATER IN THE GROUND. Small catchments, such as swales and basins, are very effective for catching water and helping it infiltrate the ground, especially if the landscape is designed to slow the water, preventing it from gaining volume or speed as it moves downhill. Small basins can catch and hold water at individual plants or plant groupings; organic mulch, like wood chips in a basin, hold the water and give it time to soak in. Start by going "upstream" to identify the sources of water coming onto the land, and then look for ways to hold the water and let it percolate into the soil. In urban areas, this might mean looking at building roofs as places to intercept and direct the water. Identifying leverage points and intercepting the water early in its flow can have a large net effect.

Rain gardens capture rainwater runoff and keep it in the soil. Channeling the water and using appropriate plants that can tolerate periodic flooding are key to the garden's success.

WAYS TO HOLD WATER IN THE LANDSCAPE

There are several ways to slow the flow of water and hold it in a landscape, so that it has time to infiltrate the soil and recharge groundwater, and to avoid soil erosion.

INFILTRATION SWALE. A shallow trench dug across a slope, on contour, to intercept the flow of water and give it time to infiltrate the ground.

BERM AND BASIN. A smaller infiltration basin on a slope with a corresponding berm downslope. Not as linear as a swale.

FISH-SCALE SWALE. A small, miniswale or minibasin set up as overlapping basins to catch the overflow of the basin above, thus the fish-scale reference. Usually used where larger swales cannot be installed or where water catchment in many small locations is preferred, such as at multiple tree locations.

CONVEYANCE SWALE. A shallow trench or channel, dug slightly off contour, to catch and direct the flow of water. The steeper it is, however, the faster the water moves, reducing water infiltration. Water bars are a conveyance swale to take water off a road and into some other structure, such as a rain garden or infiltration basin. Check dams (small dams in a stream or swale) can slow water movement for better infiltration and to even out varying flow levels. When a conveyance swale is filled with vegetation, it is sometimes referred to as "vegetated swale" or "bioswale."

RAIN GARDEN. A small to large basin filled with plants that tolerate sporadically wet conditions. Rain gardens catch and hold water and allow it to infiltrate the soil, and they can be used in any type of landscape, whether residential, commercial, or urban.

DETENTION/RETENTION BASIN. A basin meant to hold and slowly release stormwater in an effort to reduce flooding and extreme flows into streams from highly developed areas. Retention basins catch and hold water; detention basins catch and slowly release water, drying out between storms.

TERRACE. A flat planting bed on a slope that can also act as a water catchment site, slowing water and encouraging it to infiltrate the soil. Increases access and planting space as well.

FASCINE, LOG-TERRACE, OR LOG-EROSION BARRIER. Logs or bundles of sticks (fascines) staked across slopes or set into shallow cross-slope trenches. These small barriers slow water and catch sediment. Over time they can form stable microswales.

STRIP CROPPING, LIVING DAMS. Rows of perennials planted on contour can catch and hold water as well as any moving nutrients or sediment. Plants that produce a lot of biomass can hold slopes and also be used as mulch (like switchgrass in a temperate zone and vetiver grass in the tropics). Strip cropping with alternating rows of hay or seed and tillage crops like corn or beans can also catch and hold water and reduce erosion.

At the core of this rain garden – which my firm designed and installed for the Springfield Museums in Springfield, Massachusetts – is a conveyance swale, which catches rainwater from the roof, directs it away from the building, and allows it to infiltrate into the surrounding soil.

Village Homes: Effective Stormwater Management

VILLAGE HOMES, an ecologically designed community in Davis, California (see page 28), was almost never built. The process of getting the development plans approved by local government was difficult and contentious, mostly because of its innovative stormwater management plan, which was designed to keep as much water as possible on-site. The system called for moving water away from the homes and toward central open drainage channels to allow the water to infiltrate the soil. Excess water would flow into larger basins and catchments around the property as needed, depending on the volume of rainfall.

At a time when building requirements called for getting stormwater into pipes and off-site as quickly as possible, this design was controversial. The town engineer, dubious of the "radical" system, suggested declining the permit application. Fortunately, the town permitting board allowed the first phase to go ahead as planned — with the requirement that the developers post a bond to cover any costs that might result if the water system failed and caused flooding into neighboring developments.

During construction of the first phase, there happened to be some periods of intense rain, and several local areas flooded. However, the Village Homes system operated as designed, allowing stormwater to infiltrate on-site, and it even absorbed stormwater that had backed up from neighboring developments. After this successful showing, the town lifted the bond requirement for successive phases of the development project.

At a time when most housing developments funneled stormwater into the town sewer system, the Village Homes community was designed with a network of drainage channels to keep stormwater on-site and allow it to infiltrate back into the soil.

A stormwater retention basin becomes an outdoor amphitheater during dry times.

17

ENSURE
That We Have
ENOUGH
WATER

Water is essential for all life on earth. Although it covers 71 percent of the earth's surface, there is a limited supply of *fresh* water, which exists primarily in the ground (groundwater), but also in streams and lakes. Only 3 percent of water on the planet is freshwater, and a mere 0.03 percent is not in glaciers or deep groundwater. Whether you live in the drought-stricken West or in a region that experiences unpredictable rainfall, you know what a precious resource freshwater is.

Threats to our limited freshwater are largely from overuse and pollution. Groundwater levels are dropping; many water bodies are contaminated with nutrients, sediment, or toxic compounds. The future of our water supply is uncertain, and planning for our water needs should be a top priority. The ethics of permaculture — care for the earth and care for people — indicate that we should use water thoughtfully and leave the water cleaner than when it arrives to us. The ethic of fair share asks us to limit our consumption, in order to leave enough water for others. There are many ways to use less water, keep it clean, and collect the water we need for our personal and community uses.

CAPTURE RAINWATER

An astounding volume of rain falls from the sky, even in areas with low rainfall amounts. For example, a 0.2-inch rainfall (a mild storm) on 1,000 square feet of ground amounts to 125 gallons of water. This rainwater is a resource just asking to be caught and stored for use in irrigation and as drinking water.

The easiest way to capture rainwater is to collect what falls off a roof. Metal roofs, though expensive, are best suited to this type of use (after tile or slate, which are even more expensive) because they don't release petroleum compounds into the water.

Knowing how much rain falls in your area every month, and how much roof area you have, can help you gauge your rainwater harvesting potential. Even a small surface, like a shed or back porch roof, can collect large amounts of rainwater through a season. My family collects rainwater for our chickens simply by putting a bucket under a low corner of the mobile chicken coop roof; this way, we don't need to carry water to the coop as it moves around the land.

Rainwater collected off a building roof is commonly stored in a tank or cistern. Although the tanks themselves, or materials to build a cistern, can be expensive, consider it an investment in resilience. There are many shapes, sizes, and types of tanks that can fit in different sites and conditions. Rainwater barrels that hold 40 to 55 gallons are now relatively common. They are a good start and are usable on a small scale, but once your irrigation needs go beyond 55 to 110 gallons (which can happen quickly), you'll need to increase your water storage capacity. A tank that holds 300 gallons can get a small orchard through a brief drought; it's enough to give 20 trees 5 gallons each per week, over a three-week dry period.

Some tanks are designed to be set belowground. Buried tanks take up less land space and aren't a visual distraction, and depending on how deep they're set and the weather conditions, they can protect their contents from freezing. Whatever type of tank (or any water storage container) you have, set it as high up as possible, so that you can use gravity to draw water from it, without the need for pumps.

REDUCE WATER CONSUMPTION

In addition to collecting water, we can limit how much we use. We can easily take steps like installing low-flow showerheads and faucets, taking

Using rain barrels to capture water for irrigation is a simple way to cut down on groundwater usage.

shorter showers, and using an energy-efficient dishwasher instead of washing by hand. We can also address the fixture that accounts for almost 27 percent of household water use — the toilet.

Most people use a flush toilet without giving a thought to the clean water used in the process and the contamination that ensues as the waste seeps into the ground or is "treated" by water treatment plants and then flows into a nearby waterway or back into the water supply. Even a "low-flush" toilet uses 1.6 gallons per flush; using the toilet five times per day means each person is pumping and flushing 2,920 gallons of water each year. A family of four in the house may be using as much as 11,680 gallons per year just flushing the toilet. A "regular" toilet (not low-flush) uses twice as much.

THE ALTERNATIVE TO THE FLUSH TOILET IS A COMPOSTING TOILET. Some people might picture a composting toilet as a smelly outhouse in the woods, but we've come a long way since then. There are many efficient, clean ways to compost our "humanure." Systems can be simple owner-built setups or commercial units. At my home, our composting toilet is a large-batch Phoenix-brand composting toilet. After every use, we toss in a handful of wood shavings. The pile is mixed weekly with several aerating tines, and every two years we remove 20 to 30 gallons of rich, earthy compost, turning what would have been a waste product into a resource.

A popular way to compost humanure that I've used in the past is the bucket system outlined by Joe Jenkins in his *Humanure Handbook*. This is a simple, effective way to compost waste, but it requires more handling than is ideal for many people. Small batches are collected in a 5-gallon bucket and then transferred to a larger, dedicated compost pile. For larger households and for long-term use, an in-place batch composter is probably better.

Even a "low-flush" toilet uses 1.6 gallons per flush; using the toilet five times per day means each person is pumping and flushing 2,920 gallons of water each year.

An Integrated Water Cycling System

GAP MOUNTAIN PERMACULTURE in Jaffrey, New Hampshire, is the homestead of designer Doug Clayton. The water system at Gap Mountain comprises several integrated elements that work together to collect, hold, and cycle the water falling on and flowing over the land. Doug created this water system at his home to make the best use of water in a very low-energy system.

In addition, a shallow well in the orchard provides access to this groundwater, and during a drought period, a solar-powered pump cycles water back up to the cistern at the house. With these three main gravity-fed storages — the cistern, the pond, and the soil — and a solar-powered pump, the water stays on-site and makes a complete cycle through the property.

WATER COLLECTION

Rainwater is collected off the metal roof via gutters and downspouts and then is funneled down into a basement cistern. This concrete cistern, which forms part of the basement foundation, holds 3,400 gallons of water for use in irrigation and for in-home water needs. The cistern is built into the ground so that its contents won't freeze in the bitter New Hampshire winters.

Overflow from the cistern gravity-feeds into a pond downslope of the house. The pond is sealed with a layer of bentonite clay, so that the sandy, gravelly soil here can hold the water. The water of the pond and nearby large stones hold heat and create a warm microclimate, which is ideal not just for pond-side plants but also for people; it's a beautiful spot to relax. The pond also allows some water to infiltrate the soil and recharge the groundwater and support nearby plantings.

Overflow from the pond moves through a small spillway that directs the water down the slope and into the orchard below. Water flowing into the orchard infiltrates the ground, recharging the groundwater and making it available for the fruit trees and associated plants.

COMPOSTING TOILET AND GRAYWATER SYSTEM

In addition to the rainwater harvesting and cycling, Doug has a composting toilet for the solid "waste" and a graywater system for the water from sinks, shower, and laundry. The graywater flows into a 70-gallon tank inside the attached greenhouse. When the tank is full, it empties in one surge, forced by gravity, into piping that irrigates a garden bed below the house. That garden bed is filled with wood chips and planted with raspberries. The wood chips host soil organisms that take up the nutrients in the graywater, passing them to the raspberry plants. As the wood chips decompose into soil, they are replaced with fresh chips. And any detrimental organisms in the graywater become part of the food chain when they enter the soil.

The composting toilet, a design that Doug and permaculturist Dave Jacke developed (called the Gap Mountain mouldering toilet), composts the humanure in two adjacent bins used in succession. While one bin rests and its contents break down into rich earth, the other is in active use as the toilet. The bins hold solid human waste and urine as well as wood shavings or other bulk material.

FOUR WAYS TO SAVE WATER

There are many easy water-saving techniques we can incorporate into our daily lives. If we take it a step further and evaluate how we design the systems that support us, we can have an even bigger effect.

Collecting and distributing graywater (the water from washing machines and the kitchen sink) to irrigate gardens is a good way to use water more than once.

A composting toilet can save thousands of gallons of water each year.

With surface irrigation (like a sprinkler system), a portion of the water delivered is lost to evaporation; subterranean irrigation avoids this and keeps water in the ground.

When we built an addition onto our house, we incorporated a cistern into the poured foundation to capture and store rainwater.

A Self-Sufficient Homestead for "The New Normal"

FOR THE LAST FIVE YEARS, my design firm, Regenerative Design Group, has been developing an example of integrated permaculture design on an 8-acre rural homestead in Conway, Massachusetts: Wildside Gardens. It is stewarded by owner, resident, and caretaker Sue Bridge, and it contains all the elements of a fully functioning system: a small, energy-efficient, off-grid cottage with energy systems, greenhouse, root cellar, and green roof. It also includes production gardens, kitchen gardens, an orchard, a forest garden, a rice paddy, and surrounding habitats that are utilized for less-intense production.

I entered the process as the home was being built. After interviewing Sue about her intentions and spending almost a year observing the land and its patterns amid the ongoing construction, we came up with a whole-site plan. Much of the initial pattern of the site planning has remained as we designed it. But, as is the case with any design project, there have been changes as we've come to know the land better and as we've refined the project goals.

SUE'S CURRENT GOALS ARE:

1. Understand the land and steward it in a conscious manner.

2. Identify places of ecological disturbance and utilize strategies to restore and regenerate these places.

3. Establish perennial and annual food crops to enhance self-sufficiency and produce surplus to share; including winter storage crops for year-round food self-sufficiency.

4. Grow a diversity of crops and demonstrate full productivity and integration of systems.

5. Bring people on-site to see what permaculture and regenerative design looks (and tastes!) like.

The arbor is made of black locust, a rot-resistant, locally grown wood. In the background, the cottage produces its own electricity and hot water with the solar electric and hot water system, making it independent of the energy grid.

An inviting path leads through the forest garden and creates access on the steep slope.

Top:
The green roof cottage at Wildside is positioned on a south-facing slope – perfect for active and passive solar. The green roof insulates the roof and lengthens its life span. Gardens wrap around the cottage to the south for easy harvesting access and the solar hot water and electric systems are positioned out of the way to the north of the cottage.

Bottom left:
Perennials such as chives and calendula attract and feed pollinators, and can be harvested for medicine and food.

Bottom right:
The root cellar at Wildside (built into the hillside to utilize the relatively constant temperature of the earth) stores root crops and winter vegetables for use through winter and spring.

Wildside Gardens is a place of study and learning as well as production. Here, Keith Zaltzberg of Regenerative Design Group leads visitors through the forest garden.

A boardwalk moves through the wet meadow and leads to a meditation garden and wildlife pond.

A Self-Sufficient Homestead for "The New Normal," *continued*

The homestead is a place of learning and experimentation, and a way to document the development of a full permaculture-designed homestead. As such, it's become a destination for groups and classes. Sue thinks of this as an early-twenty-first-century work-in-progress, to get ready for the "new normal": a world affected by the effects of climate change.

The different areas of the Wildside Cottage and Gardens are:

1. COTTAGE AND ENERGY SYSTEMS.

A green roof, passive solar orientation, and grape arbor "eaves" help keep the home warm in winter and cool in summer. Active solar hot water feeds into a radiant floor system. The open floor plan makes the 1,050-square-foot home feel open and spacious. Solar photovoltaic panels provide electricity, and a compost toilet cycles nutrients back onto the land. Graywater is collected and utilized on-site.

2. TERRACES.
Four terraces bring production beds right up close to the house. One bed has annual vegetables, another has perennial vegetables, a third has perennial flowers as pollinator support, and the fourth has a mix of bush cherries, sunchoke, comfrey, and horseradish. Microswales above each terrace collect and distribute water across the upper side of each terrace. Additional beds around the cottage have flowers, rugosa rose, culinary herbs, and additional space for annual veggies.

3. FOREST GARDEN.
This has multiple guilds (see page 94) and various niches with different conditions. A series of three swales collects and infiltrates water. A fertility bank grows dynamic accumulators to be cut for mulch and compost making. Of particular interest are the American persimmons, aronia, and a variety of groundcovers, such as tansy, mountain mint, yarrow, and sweet cicely.

4. WET MEADOW.
On the east side of the land, a lowland with seasonal stream and wetlands creates opportunities for diversity of habitats and activity. A boardwalk winds through the meadow to reach a quiet sanctuary space and small pond. The meadow is managed for the wildlife habitat values with some plantings of production crops, using the productive conservation approach, like elderberry, basket willow, and blueberry. A rice paddy at the edge of the wet meadow creates an intensive production zone accessible from the access road.

5. NUT GROVE.
In the northern corner of the land, nut trees, which will grow large and take up significant space, are planted in an open meadow. As they establish, the meadow provides open grassland habitat. Species being grown are chestnut (both American hybrids and Chinese), bur oak, swamp white oak, hickory, and butternut. Walnuts and hazelnut are in other areas on the land.

6. PRODUCTION GARDEN AND GREENHOUSE.
A production garden along the driveway is space for larger crop production such as garlic, asparagus, potato, squash, and tomato. The earth-bermed passive solar greenhouse by the garden has a green roof. The greenhouse is the place to get seasonal veggie starts going and has room for a fig and sometimes ginger plantings. In the summer, sweet potatoes and tomatoes grow here.

7. HEDGEROW. Along the driveway, a hedgerow creates a divide and utilizes an edge space. Mature trees here are hung with bittersweet, which over time die and create snags supporting woodpeckers and cavity-nesting birds plus landing sites for hunting hawks. Hazelnut, Siberian pea shrub, hawthorn, and autumn olive are producing along here, with an area of thimbleberry, rugosa rose, and sweet fern holding a steep slope.

8. SURROUNDING MEADOW AND FOREST ZONES. Slopes are forested and managed for yields of timber while other areas are simply left to provide for the wild inhabitants. Meadows are mowed yearly to limit woody growth and supported as important open habitat.

9. WATER SYSTEM. The water system is integrated with the food production, energy production, habitat, and aesthetic elements. The system is designed to reduce the use of freshwater, to purify and reuse "waste" water, and to cycle and store rainwater.

10. FOREST GARDEN SWALES. Three swales cut across a slope to slow the passage of water and encourage it to infiltrate the soil, reducing runoff and passively irrigating the forest garden.

11. GREEN ROOF. The green roof on the house slows stormwater runoff, helps insulate the home, and provides beauty and habitat. It's a grassland in the sky!

12. GRAYWATER SYSTEM. Though it's not yet installed, the planned graywater system someday will direct graywater into mulch basins in the forest garden, where it will be filtered and available to support the food crops there.

13. COMPOSTING TOILET. The Phoenix composting toilet reduces water use to a minimum and cycles the humanure nutrients on-site. Liquids from the composting process can be collected and reused on-site as well.

14. RICE PADDY–WARMING POND. A small pond at the base of the slope collects water from the back shed roof above and brings it into the gardens. This water provides habitat and warm water to the rice paddy, which encourages rice production.

15. WET MEADOW AND POND. This is a special habitat on the property, one that provides open grassland and shrub areas for wildlife and particularly good forage for pollinators. A small pond was hand-dug in the meadow to provide open water and attracts frogs, salamanders, and other wet-loving critters.

16. BOOSTS TO SOIL WATER-HOLDING CAPACITY. Soil building with organic mulches around the property have increased the land's water-holding capacity. Now, when big storms (like Hurricane Irene) hit, the deluge of water infiltrates the soil easily, with no runoff or erosion. And the irrigation needs of the gardens and forest garden have been minimized as the soil's water-storage capacity has increased.

- -

The production gardens and earth-bermed greenhouse with a green roof allow for almost year-round production and supply a sizable amount of Sue's food. In the summer the greenhouse has heat-loving crops like ginger, sweet potatoes, and fig.

Turkish rocket is a perennial vegetable, producing yearly crops of broccoli-flower heads and edible leaves. The flowers support myriad pollinators.

1,050 square-foot cottage

Perennial vegetables

Annual vegetables

Pollinator plants

Bush cherries, sunchoke, comfrey, horseradish

Greenhouse with green roof

Graywater from the house will flow down to the forest garden.

Water warms in the pond, then runs down to the rice paddy.

Wet meadow

Remnants of an old Christmas tree farm

18

CREATE FINANCIALLY RESILIENT *Communities*

BerkShares – the local currency for Berkshire County, Massachusetts – were created as an alternative to U.S. currency, as a way to keep money circulating locally.

Money is like energy — it's most useful to us when there's a system in place to keep it moving and bring it to where it's needed, when it's needed. Like any other human-designed system, though, a monetary system is most resilient, most effective, and most beneficial to all its members when it mimics an ecosystem.

An ecosystem is a community of organisms and entities linked together through nutrient cycles and energy flows. There is no waste, and materials and energy cycle continuously through the system. The wood in trees and organic matter in soil represent reservoirs of captured energy, which in time is released and flows back through the system.

In a healthy economic ecosystem, money flows from person to institution and back again. In the current state, however, there are many one-way, linear flows in which money ends up in reservoirs and is held for long periods of time (like in an offshore account or being invested in international corporations), or it flows out of the community and is lost (where products that could be made and sold locally are purchased from faraway producers). J. Paul Getty was once quoted as saying that money is like manure; you have to spread it around, or it smells. Thinking of money like manure helps you see how unhealthy a pileup of it can be for the system as a whole. Similarly, the money lost to communities is like a farm field with its valuable soil washing into the nearby stream; a loss of capital which becomes a pollutant to the ecosystem.

A healthy financial ecosystem has many connections and interrelationships — there are lots of nodes of production, inputs of materials, cycling of capital, resources, and "waste" feeding into other production cycles. A tree grown in your area, cut by a skilled logger, milled at a local mill (or on-site with a portable mill), and made into lumber for a home or a piece of furniture by a local craftsperson, has created employment, generated a need for skilled workers, and circulated money through the community. Buying that furniture from a faraway forest cuts off that cascade of benefits. To create healthier, more financially resilient communities, we need to support and increase these interconnections and opportunities for cycling. Some ways to do this are:

- **USE LOCALLY PRODUCED** and locally grown materials and products.

- **SET UP ECONOMIC ALTERNATIVES** to traditional financial transactions, such as barter networks, buying clubs, work shares, and local currencies.

- **FIND THE LOCAL ENTREPRENEURS,** lenders, and foundations that can support a movement to expand the local business ecology, and work with them to start a local investment fund that supports business start-ups.

- **ASSESS THE LOCAL BUSINESS ECOLOGY** for opportunities and constraints in production, processing, storage, and distribution.

- **BUILD COMMUNITY** through neighborhood tool shares, seed libraries, plant swaps, and reading groups.

- **ENCOURAGE LOCAL INVESTMENT,** local business entrepreneurship, and product development using local materials and crops, and start or work with a nearby community development corporation.

- **START A FOOD HUB** and local/regional food processing center or support existing enterprises.

- **DEVELOP LOCAL EXPERTISE** in a diversity of businesses and trades, and develop programs to train youth in these trades as well as marketplace and business skills.

- **ENCOURAGE/SUPPORT AFFORDABLE ACCESS** to land and affordable housing.

- **WORK WITH LOCAL SOCIAL JUSTICE** advocates and local government to plan for local and regional food security and food justice; increasing access to healthy food and safe work conditions and living wages for food industry workers.

WHERE IS YOUR MONEY?

Where your financial capital resides is an important consideration. There are different ways to invest, and with some investments, the return is not a boost to your financial wealth but a boost to your own and your community's long-term health, stability, and resiliency. Is your wealth in a large corporate bank, global investments, or gold and jewels? Or is it in a local credit union or socially and environmentally responsible investments? Is some of it invested in land, orchards and gardens, water or energy systems, livestock, or a renovated home or other structure?

If you want to support your values with the financial resources you have, consider where your money is, and take steps to make sure that it supports well-being, justice, and resiliency for your community and the world at large.

Thinking of money like manure helps you see how unhealthy a pileup of it can be for the system as a whole.

Preserving life should be the natural result of commerce, not the exception.

～～～～

— Paul Hawken

BEYOND FINANCIAL CAPITAL:
WHERE DOES YOUR WEALTH COME FROM?

Money is just one way of assigning value to the things we own or that are held within our community. In the article titled "Eight Forms of Capital," Ethan Roland and Gregory Landua outline the various ways we can think of capital beyond the purely financial sense. All of these forms of capital come together to enrich our lives and to connect us more deeply to our families and communities. In fact, true resilience and sustainability might rely more on these other forms of capital than on our financial wealth. In addition to financial capital, the other types of wealth are:

SOCIAL CAPITAL:

Influence and connection

MATERIAL CAPITAL:

Nonliving, physical objects

LIVING CAPITAL:

Animals, plants, water, and soil

INTELLECTUAL CAPITAL:

Knowledge, understanding, and the ability to solve problems and design solutions

EXPERIENTIAL CAPITAL:

The understanding that comes from doing work, planning and implementing projects, and gaining wisdom through direct experiences

SPIRITUAL CAPITAL:

Connection to self and the wider universe; personal and profound, and difficult to quantify

CULTURAL CAPITAL:

The shared knowledge, understanding, practices, and traditions held by a community

19

REVITALIZE NATURAL AREAS

while Providing for Humans

Our collective awareness of environmental issues has grown in recent decades as we have borne witness to the destruction that humanity has caused on the planet: loss of habitat and biodiversity, the unraveling of whole ecosystems, the erosion of topsoil, the flow of fertilizers from farms into the Gulf of Mexico, the creation of giant anaerobic "dead zones" in the oceans, where fish and other creatures cannot survive, and on and on.

While it's become clear just how much damage humans are capable of doing, unfortunately, we've come to see ourselves as only the source of environmental *problems*, not as the source of healing. The environmental movement evolved with a deeply embedded attitude of separation — humans are separate from nature, a defective and dangerous species (and also superior at the same time). The conservation movement has internalized this separation as it seeks to protect land and species from destructive human forces. Understandably so, since for the last two thousand years (and probably since the rise of agriculture some 15 thousand years ago) we have occupied and manipulated more and more land and sequestered much of the planet's resources for our needs.

But by perpetuating the idea that separation is necessary to prevent destruction, we're not recognizing our ability to be a positive force on the land. In this perception, natural areas are places without people, and developed areas have no nature worth protecting or investing in. This idea — that we have "pure" nature on one side, which we must protect, and human society on the other, where we plunder as we wish — fails to account for the potential of humankind to reweave ourselves and our infrastructure into the network of natural ecosystems around us.

TENDING THE WILD

An alternative paradigm to "nature versus human" is the idea of humans as stewards of the earth. This isn't a new idea. In many cultures before our modern industrial society, humans played an active role in managing the landscape, shaping the development of local ecosystems, and, as author Kat Anderson so eloquently puts it, people have, for time immemorial, been "tending the wild." Indigenous peoples worked within the bounds of wild landscapes to provide for their own needs while also supporting other animal and plant species, as well as local watersheds and soil ecologies, through sophisticated land management techniques such as fire, game management, spreading seeds of desired plant species, and pruning. Indigenous "gardens" may not have looked like what we think of as a garden today, but those wild-looking landscapes were, in fact, carefully cultivated for health, diversity, and production.

Early accounts of Europeans arriving on the eastern seaboard relate how the open understory allowed horse riders to move through the forests with ease. The forests were open below, with towering, cathedral-like tree canopies, primarily as a result of the native tribes' seasonal practice of setting fires to clear the forest floor, supporting game, regenerating tree and shrub growth, and suppressing pest and disease organisms. But quickly after European contact with the New World, virulent diseases spread through the Native American communities and wiped out as much as 90 percent of the population. The Native Americans could no longer manage the land, and the forest filled in with thick growth, reducing habitat for wild game and fundamentally changing the characteristics of the land. The European colonists who arrived in later years saw the forest as a dark and dangerous place, thick and tangled, and they undertook to cut it down and replace it with "proper" cultivated land.

The idea that humans are separate from nature (indeed, that humans are a danger to nature) continues to this day, making it difficult for people to see themselves as part of the landscape, with the potential to help heal the land. But we have the potential to initiate a tremendous revitalization of natural places, simply because nature itself has huge potential for regeneration. Think of the northeastern forests: though fairly well decimated in the 1700s and 1800s by logging and agrarian clearing, once the land became more sparsely settled and many farms were abandoned, the forests grew back. About 80 percent of the Northeast is now forested — again, and many of the wild animals like beaver, moose, and bobcat have returned.

The land regenerates. And it can do so faster, while supporting relationships and providing benefits to all manner of flora, fauna, and ecological niches, including humans, with some help from us. All it takes is a willingness to consider living on the land in new ways.

PRODUCE FOOD AND REGENERATE THE LAND

"Productive conservation" is the practice of both protecting and conserving land while also growing crops using low-intensity agricultural practices. In this pattern, human activity on the landscape improves the vitality, functionality, and beauty of natural areas, open space, and even sensitive habitats. We become active participants, working the land for our needs and goals while also stewarding the land.

With productive conservation, the most intensive, high-disturbance practices happen in the least environmentally sensitive areas. Around these areas are productive conservation zones, where some low-disturbance agricultural activity takes place but the focus is on maintaining ecological function. These zones surround the places we want to protect: sensitive ecosystems, habitat for rare species, and special places on the land, such as sacred sites, places with a view, public areas, homes, and wherever we want some buffering from intensive land uses.

Often floodplains and riparian zones (the edges of waterways) — where the needs of farmers and ecosystem function can often be at odds — are good candidates for these approaches. Floodplains are important ecologically for recharging the groundwater, filtering sediments, slowing erosion, and providing wildlife habitat, but they also tend to be very fertile, and prime ground for agriculture.

Here's an example of how this apparent conflict can be overcome. Imagine you're walking along the river, through a woodland with towering cottonwood and sycamore trees, with a smattering of nut trees such as northern pecan, oak, and hickory. Birds sing high up in the trees, and

The practice of "productive conservation" protects sensitive ecological zones, like this riparian area, and limits intensive agricultural production to less ecologically sensitive areas.

animal tracks are scattered across the mud and sand. Shrubs, including elderberry and blueberry bushes, are thick under the trees, and in sunny openings and woodland, herbs like American ginseng, goldenseal, and black cohosh carpet the forest floor. It's a healthy forest, keeping the river in cooling shade, providing habitat and food for wildlife and pollinators, and filtering nutrients and sediment headed toward the river from neighboring farms. But it's not a *wild* forest; it's semi-cultivated. People have planted the nut trees, berries, and medicinal herbs, harvesting them carefully with minimal disturbance, and at the same time the plantings are contributing to a healthy ecosystem.

This is one way to put productive conservation into practice, but of course there are many others. Some principles include:

- **USE TREE CROPS** and deep-rooting perennials to get yields while stabilizing the soil.

- **MAKE USE OF MULTIPLE VEGETATION LAYERS** (ground, shrub, tree, vine, root) and polycultures of multiple species to diversify productive niches.

- **UTILIZE PERENNIAL CROPS** such as medicinal herbs, perennial vegetables, nuts, berries, mushrooms, fiber, fodder, and biomass.

- **FOCUS ON LOW-INTENSITY** cropping with minimal interventions, and dispersed yields in the most sensitive areas.

- **STAY OFF STREAM BANKS,** let them remain thickly vegetated, and let streams and rivers meander; adjust your productive conservation planting with the river movement.

- **USE PRODUCTIVE CONSERVATION** agricultural activity to enhance ecosystem services; slow and infiltrate water with vegetation, swales, or basins; leave woody debris to reduce erosion, provide habitat, and build soil; use coppice and deep-rooting plants to sequester carbon.

- **AVOID SOIL EROSION** by using minimal tillage (or no-till methods) and keeping the soil vegetated.

- **PLANT HEDGEROWS,** and diversify the edges of fields and landscapes with plantings that yield a crop while they support pollinators, reduce wind erosion, and provide other benefits.

- **DON'T ADD FERTILIZERS,** raw manure, or herbicides and pesticides to productive conservation areas near shallow groundwater, surface waters, or sensitive habitats.

- **DON'T BUILD ROADS** in areas near rivers, wetlands, or sensitive habitats.

- **BUFFER VERY SENSITIVE HABITATS** with vegetation that will, at most, be only minimally disturbed.

We have the potential to initiate a tremendous revitalization of natural places, simply because nature itself has huge potential for regeneration.

INVASIVE SPECIES: NO SIMPLE ANSWERS

These days, so-called invasive species are much maligned in the press and in conservation circles. The concept of species "invading" and taking over ecosystems is relatively new, however; it became a focus of environmental science since the 1950s. New species, spread by human activity, are certainly changing the ecosystems they merge into. In rare cases, primarily isolated habitats, they can indeed reduce or eliminate populations of native species. But in the vast majority of cases these newly arrived, or naturalized, species become part of the local ecosystem processes and successions, eventually becoming indistinguishable from species that arrived 50 years ago, a hundred years ago, or even a thousand years ago.

Examining the topic of invasives through this kind of lens helps give some perspective to a divisive issue – one that is fraught with warlike attitudes. At one point in my own career as a restoration ecologist for the Nature Conservancy, I battled invasives with this frame of mind, and in some cases with large quantities of Roundup discharged from tractor-mounted spray equipment. I'm proud of the land we restored, but I'm not proud of the practices we used and the state of mind that the work engendered. The "us" and "them" attitude only serves to separate us from the other players in the ecosystem we're a part of.

These days, I'm more interested in finding ways to regenerate the ecosystem without going to war with it. In fact, scientists are beginning to understand the ways in which invasive plants themselves may actually play a role in regenerating an ecosystem. Habitat destruction and disturbance by humans create opportunities and clear the way for tough, early succession species. It turns out that many of the so-called invasives excel at ecosystem restoration, by filtering sediment, taking up nutrients, holding back soil, and providing habitat.

Japanese knotweed

The salt cedar tree provides just one example of this. After generations of humans clearing vegetation along rivers in the desert Southwest, withdrawing water, and over-grazing livestock, the ecosystem is highly changed and, in many ways, damaged. One newly arrived species, the tamarisk or salt cedar, is drought and salt tolerant and is taking advantage of the changed conditions to grow extensively along stream corridors. Despite its reputation for lowering the water table, rais-ing salt levels, and taking up more water than native plants (an accusation that has since been disproved), the tamarisk has also proven to be the preferred nesting site for the endan-gered southwestern willow flycatcher. After decades of maligning and destroying the tama-risk, we must now reconsider our stance if we want to help the flycatcher. The tamarisks, and other newly arrived species, may just be the ecosystem's response to our activity, trying to mend the torn fabric of the land.

So we are left wondering: In such complex systems, under such massive change, how do we know what kind of action we should take to "save" ecosystems? The simple answer is that we should stop the intensive interventions and spend more time trying to understand what is happening, and why. Big decisions made with intense emotions, lacking solid observation and scientific research, can lead to bad results. Take the example of the biologists in the Pacific Northwest who decided in the 1960s and 1970s that debris in rivers was the cause of salmon decline. This led to the removal of debris jams in rivers across the region, which led to mas-sive erosion and loss of streambed habitat, which happened to be essential for salmon

Purple
loosestrife

spawning and the macro-invertebrates salmon eat. That intervention, undertaken to help save the salmon, actually led to worse conditions.

Could we be doing the same thing in our quest to cleanse ecosystems of "bad" species? Probably. Scientists are beginning to gather evidence that these species are actually helping restore habitats and ecosystem function. From purple loosestrife and zebra mussels to kudzu and Japanese knotweed, we are learning that these species are finding their places in "novel" ecosystems and becoming part of the solution to habitat disruption. Scientists are studying these novel ecosystems and finding that these species might be the partners we need to fix what we've broken.

Kudzu

20

Turn Problems
INTO
SOLUTIONS

Limited growing space could be considered a "problem," or it can be thought of as an opportunity to come up with creative ways to maximize growing space and enrich the soil to boost productivity.

When we're facing a simple problem, we attempt to fix it directly. The sink is leaking, so we get a new gasket and repair it. But with more complicated problems — especially environmental problems that affect many people with different goals — we might find that the solution is not so easy to find.

One tactic is to take a step back and look at all the aspects of whatever situation you're dealing with. Within that problem, there might be a solution that goes to the core of the issue. For example, if erosion is causing sediment to flow into a stream, removing that sediment undoes the damage, but it doesn't solve the problem. The solution might entail going upstream and slowing the water before it picks up the sediment, or creating a filter strip of vegetation along the waterway to take out the sediment. You might even be able to turn the problem into a solution; in a small, eroded stream, for example, you could trap the sediment with a small dam (a "check" dam) to help rebuild the streambed.

A classic example of the problem actually being the solution is wastewater. This "waste" flows out of our homes and into septic systems or waste treatment plants, costing money, using freshwater resources, and even polluting waterways and groundwater. But there's actually no need to treat this material as waste; rather, we can see it as a resource too valuable to waste. The graywater can provide supplemental irrigation water, especially in areas with low rainfall or drought conditions. The solid waste can be composted and recycled onto our properties or utilized in nearby farmlands. Urine, with its high levels of nitrogen, can be cycled onto agricultural lands or anywhere fertility is limited. The problem of waste is really an opportunity.

Look at the problem, and then look for an opportunity nested within it.

PROBLEM: The compost pile is too far away.

OPPORTUNITY: Moving the compost closer to the house will allow you to more actively use it, boosting the amount of compost available for your garden.

PROBLEM: Animal manure or fertilizers are polluting a waterway.

OPPORTUNITY: The manure or fertilizers could be captured and used to fertilize a "wasteland" environment to grow beautiful flowers or plants to support pollinators.

This perspective or "paradigm shift" can lead to searching where we live, for solutions big and small. The challenges we face sometimes need fresh out-of-the-box thinking. Sometimes a problem that seemed large and insurmountable can become an easy, elegant fix, if we see it another way.

The problem of waste is really an opportunity.

21
FULFILL OUR ENERGY NEEDS

We are addicted. Fossil fuels, from the coal mines of the industrial revolution to the oil and natural gas that power the infrastructure of our modern economy, have transformed our society so completely that we have lost sight of how dependent we are on them.

Globally, 84 percent of the world's energy comes from fossil fuels. This seemingly abundant, cheap energy source powered amazing developments in agriculture, infrastructure, technology, and transportation, and it has been an important factor in the exponential growth of human population. We wouldn't be where we are without it.

As we've been learning more and more over the last several decades, burning fossil fuels and the resulting release of carbon dioxide has contributed to global warming, which is primed to cause potentially catastrophic climate change all over the world. For this reason alone, it makes sense to find other ways to fulfill our energy needs. But whether or not that fact motivates us to make a change, we'll eventually be forced to seek other energy sources, because for years now, we've hovered at the edge of an energy precipice: peak oil.

Oil availability globally is peaking. Though oil and natural gas from tar sands and fracking have been able to shore us up over the past few years, the technology behind their extraction is fraught with controversy about pollution, damage to the local geological structure, and cost. There's no denying that we are in a very precarious place right now, where discoveries of new mega oil fields are in steady decline, while our consumption of these fuels is only increasing. As this gap widens and energy prices increase, disruptions to every aspect of modern life will ensue. The time to act is now (or decades ago, but since that time is past, let's move on).

Fossil fuels are inherently unsustainable because they are not renewable. After all, it took millions of years to produce them; we can't wait that long for the cycle to come around again. It will not be easy to transition to renewable sources, but we must. What to do?

REDUCE CONSUMPTION

First, we can reduce our energy consumption. In the United States, we use a *tremendous* amount of energy. With just 5 percent of the global population, we consume more than 20 percent of the world's energy. It wouldn't make sense to scale a renewable energy system to current uses before implementing intensive conservation and energy-reduction measures. There are many ways to reduce our use of energy, as a nation, as communities, and as individual households. In our own homes, we can use energy-efficient appliances, reduce our use of appliances that consume a lot of energy (like air conditioners, clothes dryers, and electric space heaters), and so on; there are lots of good tips on saving energy from sources such as the U.S. Department of Energy. On a community level, we can support public transportation. Invest in bike lanes and paths, to support commuters using pedal power rather than their gas guzzlers. We can switch street and traffic lights to LED and retrofit municipal building to use less energy. Towns and cities can switch their vehicles to electric or very high efficiency vehicles. And as a nation, we can give support to renewable energy sources, get utilities to offer (or continue to offer) energy efficiency incentives to customers, fund energy efficiency programs, and make investments in energy efficiency at all levels of the energy supply chain. Building codes often need to be changed to reflect energy efficiency goals and to continue to reduce energy use in appliances and increase use of new, effective insulation materials. Some states have set energy efficiency targets and established stable (multiyear) funding for energy efficiency programs. Others can follow this example.

One example of a community taking action to reduce their energy consumption occurred as a result of a crisis. In April 2008, an avalanche cut hydroelectric power transmission lines to Juneau, Alaska, the capital city with a population of 31,000 at the time. This corresponded with a spike in diesel fuel prices. Facing a dramatic increase in electricity bills, the residents implemented intensive energy-saving measures; mostly changes at home such as line-drying clothes and changing to compact fluorescent bulbs. These simple changes led to a 40 percent drop in electric usage in just six weeks! The capital city went from peak power usage of 50 megawatts before the avalanches to below 30 megawatts by late May. Given some incentive, we can make these changes.

MAKE YOUR OWN ENERGY AT HOME

In permaculture, we aim for a level of self-sufficiency in our communities and our homes that allows for control over our choices and keeps our needs taken care of close to where we live. With many people producing energy dispersed over a larger area, the system has redundancy, an important principle, and allows local control so any disruptions don't affect the entire area.

What energy can you produce? Assess your potential for renewable energy generation at home. If you are lucky enough to have a stream on your land and some topography (a drop in elevation), you might be able to harness micro-hydropower. Water runs all day and night, and even a small water supply can provide enough power for a home or neighborhood, 24/7, 365 days a year if the stream runs year-round. Microhydro comes from proven technologies first developed in the late nineteenth and early twentieth centuries and changed very little since then. Costs for systems are reasonable if the conditions are right. The primary cost is for the pipe to direct the water (under pressure) to the electric generator.

My wife, Kemper, and son, Forest, generate pedal power electricity at the Real Goods Solar Living Center in Hopland, California. Pedal power can be used to generate electricity, pump water, grind grains, and power tools and equipment.

Wind turbines generate significant amounts of electricity on western rangelands.

Wind power can be a great option in some high-wind areas like ridges, the coastline, and the Midwest plains. Smaller, home-scale wind turbines are effective but require a tower to get the turbine above any surrounding trees and up into the air where the wind is less turbulent and moves quickly and more steadily. As is the case for most power generation systems, there are big differences in the scale of wind generation, with large-scale systems having turbine rotors that are over 50 meters long. Home-scale systems have rotors more in the range of 15 meters long and can be placed on towers that are 30 meters or more high.

Solar-powered electric and hot water systems are available to many people as long as they have good sunlight and a suitable location for mounting a system. Ground-mounted systems take up quite a bit of space, and they make sense only where land is plentiful or they can be combined with other uses, such as providing shade for a parking area or being set on land that doubles as grazing pasture (if the panels are fairly widely spaced). Roof-mounted systems are often an excellent choice because they usually offer excellent sun exposure. Newer metal roof systems like standing seam can allow you to mount solar panels without puncturing the roof. Prices on solar photovoltaic systems are coming down, and many states have a rebate program to make the systems more affordable.

COMMUNITY POWER

A big issue in energy supply is where the control is: who runs the energy production operation, who is getting hired, who owns and is getting the profits from energy production? From a permaculture perspective, it's important for energy benefits and money to stay within the communities where the work is done and where the energy is produced.

Deregulation, one option for local control, often leads to short-term price reductions but no lasting changes to the underlying system. Publicly owned and cooperative utilities provide lower cost electric service, according to a study by the American Public Power Association in 2015. Municipalization, where the city buys the utilities' hardware and maintains and operates the energy grid and becomes the supplier of energy, offers cities local control and the opportunity to innovate in services and reduction of carbon output and support of renewable energy sources. This is being done in Boulder, Colorado, by Austin Energy in Texas, and by Gainesville Regional Utilities in Florida with some success. Although in the Boulder case, the city and utility are negotiating the cost for the transfer and there is a potential for the case to be litigated.

An example of community control and ownership of power generation is Co-op Power in western Massachusetts. This consumer-owned cooperative provides financial and technical support to members for renewable energy systems, energy efficiency, and community energy projects and jobs. They support Northeast Biodiesel, which is in a multiyear process of building a biofuel production plant that will convert waste oil to "biodiesel." The community solar program provides financial support for residences and businesses to install solar hot water and solar photovoltaic systems. With the Neighbor to Neighbor program, members can help install others' systems and then get community help in their own installation; a modern adaptation of the New England barn raising!

**With just 5 percent
of the global population,
Americans consume
more than 20 percent of
the world's energy.**

Inspired by the chemical structure of carbon, Xome Architects in Mexico City designed the London Tower Farm, a visionary and fully self-sufficient apartment building that incorporates rainwater collection, energy generation, and a vertical farm.

22

BUILD A RESILIENT FUTURE

When the power goes out in your town — perhaps because a storm takes down power lines — what do you do? For many of us, our food spoils (refrigerators don't work), we can't flush our toilet (well pumps operate on electricity), we can't use land-line phones, and we lose our heat. Life as we know it comes to a standstill, and we realize that our lives cannot handle this bump in the road. If the power outage goes on for more than a few days, our families and communities begin to have many other problems, with transportation, communication, food supply, water supply, and even disrupted medical care.

Like a marble rolling in a bowl, our lives, the systems we create, and entire ecosystems are constantly being pushed out of equilibrium by disturbances. Something shifts the bowl, and the ball slides and rolls around rapidly, eventually coming back to equilibrium at the bottom of the bowl, at least until another disturbance comes along. This constant shifting and responding is called dynamic equilibrium.

You could say that resilience is a measure of a system's ability to maintain dynamic equilibrium — to absorb a disruption and return to balance — and of the time it takes to do so.

For humans, on an individual level, disturbances to our equilibrium might be a power outage, a death in the family, a job loss, or a storm that causes damage to the home. On a community scale, they might include contamination of groundwater or soil, drought, economic downturn, or population loss or expansion. Ecosystems, in turn, can be disturbed by changes in nutrient flows, species composition, or weather conditions.

Beyond a certain limit, a system cannot recover from a disturbance. Once that threshold is reached, the marble leaps out of the bowl; the system is changed and a new equilibrium is reached. In a land deforested so many times that it cannot support vegetation, the water table may drop, and this cascade of effects leaves it in a new state: a scrubland or desert. The land has adapted and found a new equilibrium, but the new state doesn't look anything like it did before.

Building strong local networks makes communities stronger and more resilient.

This is the concern of our times: how to adapt to a severely altered world, in which whole ecosystems face collapse, our rainfall patterns are unreliable, storms may be bigger and/or more frequent, and our lives in general are vulnerable to unforeseen disturbances. Preparedness, self-reliance, a more localized and adaptable economy, health and well-being, and a vibrant community — these are the measures of resilience for our future.

We cannot stop the changes that may be coming, but we can get prepared. We can cultivate resilience in our homes, our lives, and our communities. We can learn skills for self-reliance and teach them to each other; we can strengthen economies by buying from local stores and farmers, and by operating outside the monetary system with bartering and sharing. We can grow some or all of our own food and support agricultural biodiversity by growing or buying many different kinds of foods. We can strengthen our community infrastructure with diverse and dispersed sources of energy and water; we can invest in tools, knowledge, and local resources to bolster our own and our community's resilience.

COMPLEXITY AND RESILIENCE

As our communities and systems become more complex, the potential for disruption increases. At the same time, globally, the number of catastrophic weather events is increasing. The ability to respond and bounce back from crises requires enormous effort and coordination between organizations, governments, and communities. And the cost of response, after the crisis, is usually many times more than the cost of preparedness would have been. The cost of reinforcing the levees around

RESILIENCE: The ability of systems to respond and adapt under difficult, changing conditions.

New Orleans against hurricane storm surges, for example, would have cost 10 times less than rebuilding the neighborhoods after Hurricane Katrina.

Get people involved and excited about the changes they can make. Strategies to build resilience in your community are as diverse as the people in the community. Here are some ideas:

FOOD, WATER, AND PRACTICAL SKILLS:

- **LEARN THE PRACTICAL SKILLS** involved in growing food, like saving seeds, growing vegetables, managing an orchard, food preservation, and animal husbandry.

- **EXPAND LOCAL RENEWABLE ENERGY** generation with solar, wind, and microhydro systems.

- **BRING IN PLANT AND ANIMAL** agricultural diversity from farther away. Look for species, breeds, and varieties that are adapted to potential future conditions.

- **BREED AND SELECT** for diversity in plants and animals.

- **PROTECT LOCAL** and regional freshwater resources.

- **ENCOURAGE WATER** to infiltrate the ground wherever possible to recharge aquifers.

- **GROW FOOD** wherever you can: on public land, commercial land, prison grounds, golf courses, college campuses, empty lots, rooftops, backyards, and patios.

COLLABORATION AND COMMUNITY:

- **SET UP A TIME BANK** where members trade their skills and time with others. Instead of direct barter, a time credit system helps the flow of exchange.

- **SET UP A SKILLS EXCHANGE** in which the sharing of resources and knowledge builds networks of connection through your community.

Encouraging green space and food production in every part of the urban environment will help make cities more livable.

- **DEVELOP A TOOL** or seed library, where people can stockpile seeds and tools to lend out to other community members.

- **ORGANIZE COMMUNITY GATHERINGS** to build a strong network.

- **DEVELOP LOCAL FINANCING TOOLS** through banks, credit unions, and local foundations. Tap into or set up a Slow Money network to support local food system investment.

- **HOST A "PERMACULTURE BLITZ"** to remake a property (private or public) with food, water, and energy systems, and seek community input about how to do these things. You'll never know about the skills and expertise of your neighbors unless you give them a chance to put them to work!

SYSTEM CHANGES:

- **WORK WITH COMMUNITY ORGANIZATIONS** and policy makers to prepare an "energy descent action plan" with the steps your community can take to move toward a resilient, lower-energy future.

- **START A TRANSITION TOWN** group to build community support for addressing climate change and doing resilience work.

- **WORK WITH COMMUNITY ORGANIZATIONS** and policy makers to develop a regional or local food resilience plan, assessing the availability of agricultural land, new farmer training, farmers' needs, and production and distribution opportunities and constraints.

- **RAISE MONEY** in your community to protect good agricultural land from development and make it available to farmers in need of land.

In our own community, several friends put the call out to the community to gather and talk about permaculture, to discuss what we could do to get more people involved and more projects going in the area. About 20 people showed up, and the conversation was energetic and excited. Just being together, talking about permaculture, sowed the seed for more connection and engagement. This small event led to the establishment of the Permaculture Guild of Western

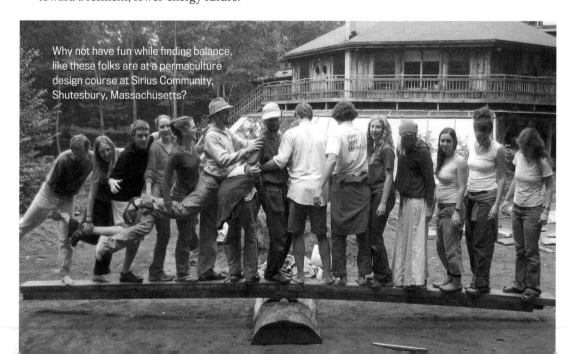

Why not have fun while finding balance, like these folks are at a permaculture design course at Sirius Community, Shutesbury, Massachusetts?

Massachusetts — an essentially ad hoc organization with very little structure, meant to support the local network. The result of this organizing effort has been a host of activities over the last 10 years, including movie nights, book discussion groups, events calendars, tool care workshops, plant and seed swaps, and a lively email discussion group. The simple act of getting together with neighbors has led to a cascade of positive connections and shared learning.

There are so many ways we can change our lives and work for change in our communities. The future is uncertain, but that doesn't mean we can't work toward positive change. Act on your strengths and help others utilize theirs. Building a better world will take a lot of time and effort, but that first step — taking action — is not difficult. This is our time, and this is the opportunity to be the change we want to see in the world.

These Flowforms at the Real Goods Solar Living Center in Hopland, California, direct water down through a series of small pools where the oscillating water is mixed and energized.

BASIC PRINCIPLES OF PERMACULTURE SYSTEMS

Permaculture design and action has long been guided by a distinct set of principles – an abstract concept that allows us to think across scales, locations, and cultures. We use these principles to understand our systems, create our plans, guide our actions, and evaluate our results.

The permaculture principles below are not meant to be all-encompassing or comprehensive; we've included the ones we feel will be the most useful to you in your daily activity and practice. These principles have their roots in many different sources, and they have been edited, added to, and adapted over time as we have come to better understand them. These principles of permaculture are living concepts that you too can implement and refine.

Harness Awareness, Intuition, and Intention: Principles of Intervention and Design

OBSERVE, ACT, OBSERVE AGAIN. First look, then do. Take a step back and observe a system or situation to discover the patterns and processes underway, and then identify the best places to intervene. For example, if you have issues with water drainage in a landscape, go out in a rainstorm and follow the flows. You may be surprised to find that the water behaves differently than you had anticipated. Perhaps try moving some soil to get the water to flow a different way, just to see what happens; let's call this "active observation." In other words, intervene in the system in order to make better observations. Act, then observe. It takes time and attention to gain deeper clarity of a system or situation. Sit and take in the surroundings or move slowly around a space. Pay attention to all the facets of a place, all the little clues. Do this observation over time in different seasons and conditions.

USE YOUR IMAGINATION. As Einstein rightly said, "Imagination is more important than knowledge." Imagination as the ability to perceive what is not visible. Use it to consider what may have happened in the past on a site and what might happen in the future. Use your imagination to think about what is going on underground. The possibilities of a system are theoretically unlimited. The only limit on the possible number of uses of a resource is the amount of information and imagination of the designer. But imagination without knowledge is fantasy. In other words, be sure to test your theories as much as you can before making large or important changes.

USE SMALL AND SLOW SOLUTIONS. Start with small solutions simply because they are manageable at the human scale. Start slowly because this allows you to harness more of the natural processes. A small and slow start also allows you to intervene and change course if things go awry. For example, planting seeds for 100 trees will yield far more, in the long term, than planting one big tree. This also allows the small trees to adapt to their place over time. And if it turns out to be the wrong kind of tree altogether, then 100 seeds is a smaller investment than one large tree.

CREATE SMALL-SCALE, INTENSIVE SYSTEMS. Start small and create a system that is manageable and produces the highest yield possible. When you start small, your mistakes are small and your learning curve is less steep.

MAKE THE SMALLEST CHANGE FOR THE GREATEST EFFECT. The less change that is made the less embedded energy is used to create the system. Leverage your resources.

CREATE SELF-SUSTAINING SYSTEMS. The role of successful design is to create a self-managed system.

USE APPROPRIATE TECHNOLOGY. What is appropriate in one context may not be so in another. This principle applies for cooking, lighting, transport, heating, sewage treatment, water, and other energy needs.

Start with Nature and Build on What's There: Principles of Ecology

WORK WITH NATURE. Living systems – such as those that support plants, animals, and ecosystems – have their own inherent patterns and processes. By utilizing or designing around these patterns and processes, we are able to create low-input, self-maintaining systems. What we have to remember, though, is that when we intervene in a system, we can't completely *prevent* a natural process from occurring. That action, or process still has energy and needs to occur in some way. In order to maintain a functional system, the intervener needs to provide a way for that function to still occur; in permaculture, we call this "shifting the burden to the intervener." For example, wild animals won't stop moving from one territory to another just because we build a freeway between them. In order to function more like a part of the system, and to avoid traffic accidents, it's a good practice to provide a wildlife corridor across the road to allow for their movement.

WORK WITH AND GUIDE SUCCESSION. Succession is the pattern of change in ecosystems over time. Planning for succession rather than fighting it is essential for ensuring resilience. Expect change and consider multiple time horizons.

CONSIDER THE EDGE. Optimize the "edge effect." Edges, or ecotones, are areas where two ecosystems come together to form a third, which often has more diversity and fertility than either of the original two. For instance a forest-meadow edge will, in some cases, support a greater diversity of plant and wildlife species than either of the adjacent areas, because it contains aspects of both.

Cultural edges are important as well. Edges are where new ideas, creativity, and change often come from. Changes in thinking about our relation to nature began at the edge and become more accepted over time as they mature.

Make Connections and Integrate: Principles of Function and Relationship

DESIGN WITH MULTIPLE FUNCTIONS IN MIND. Choose each element in a system and place it so that it performs as many functions as possible. For example, a pond creates a cooler microclimate; supports ducks, fish, and aquatic plants; catches rainfall, which can be used later for irrigation (and the clay dug from the pond can be used for building). A berry hedge serves as food, fence, wildlife forage, and nectar for bees to make honey.

BUILD REDUNDANCY INTO OUR HOMES AND LIVES. Support every function by multiple elements. Planned redundancy ensures that all important functions will be met in the event that one or more elements fail. Examples include planting a diversity of food crops and drawing on a number of different energy sources.

CREATE FUNCTIONAL INTERCONNECTEDNESS. Functional interconnection arises when the natural products of one community member meet the needs of another community member – one person's waste is another's treasure. This reduces work, waste, and pollution, since the system meets its own needs, and the natural products of each member end up as a resource for another.

LOCATE ELEMENTS IN RELATION TO EACH OTHER. Recognize connections and locate elements in relation to each other, in order to maximize their functional relationships. Each component is part of a whole, and the interconnections between them are as important as the particular components.

ACCEPT AND INCREASE DIVERSITY. As sustainable systems mature, they become increasingly diverse. Diverse systems are resilient through the myriad interconnections they make.

FOCUS LOCALLY. Grow your own food. Create community wherever you live.

"STACK" FUNCTIONS. There are two ways to stack functions in a design. One is to use elements that have multiple functions, often referred to as "stacking functions." The other is vertical stacking – utilize vertical space, such as trellising, espalier, multiple vegetation heights, and fences and walls.

Harness Energy Moving through a System: Principles of Resources and Energy

OBTAIN A YIELD. Create productive ecosystems that take care of local needs. Production of energy, food, medicine, construction materials, biomass, habitat, and others should be woven throughout communities and the ecosystem.

UTILIZE BIOLOGICAL RESOURCES. Living things grow and reproduce, building soil, contributing organic matter, and expanding the complexity of life on the planet. Protect and utilize this biological intelligence to produce food, medicine, and energy.

INCREASE ON-SITE STORAGE. The work of the permaculture designer is to maximize useful energy and resource storage in any system – be it the house, livelihood, urban or rural land, or garden. For example, store water high on the landscape where it has the most energy, or store it in soil (see page 101) where it is the least expensive to keep. The storage capacity of your system is like a bank account: the more you have in your account, the easier it will be to tap into during difficult times. Moreover, any resource that you can store when it is abundant will create more possibilities for future yield.

USE ON-SITE RESOURCES. Determine which resources are available and entering the system on their own, and then maximize their use. For example, keep rain or creek water on-site with a cistern or water tank so that you can irrigate without needing to pump from town or well water.

UTILIZE RESOURCES AND ENERGY FOR THEIR BEST USE. For example, clean water should be used for drinking and cooking, not flushing toilets.

UTILIZE ENERGY FOR LONG-TERM NEEDS. Invest in long-term systems that produce yields over time and reduce maintenance.

RESOURCES

General Permaculture

Alexander, Christopher, Sara Ishikawa, and Murray Silverstein. *A Pattern Language: Towns, Buildings, Construction.* Oxford University Press, 1977.

Holmgren, David. *Permaculture: Principles and Pathways Beyond Sustainability.* Holmgren Design Services, 2002.

Jacke, Dave, and Eric Toensmeier. *Edible Forest Gardens, Volume One: Ecological Vision and Theory for Temperate Climate Permaculture.* Chelsea Green Publishing, 2005.

——. *Edible Forest Gardens, Volume Two: Ecological Design and Practice for Temperate Climate Permaculture.* Chelsea Green Publishing, 2005.

Klein, Naomi. *This Changes Everything: Capitalism vs. the Climate.* Simon & Schuster, 2014.

McHarg, Ian L. *Design with Nature.* Natural History Press, 1971.

Merkel, Jim. *Radical Simplicity: Small Footprints on a Finite Earth.* New Society Publishers, 2003.

Mollison, Bill, and Reny Mia Slay. *Introduction to Permaculture.* Tagari Publications, 1991.

Mollison, Bill. *Permaculture: A Designers' Manual.* Tagari Publications, 1988.

Nearing, Helen, and Scott Nearing. *Living The Good Life: How to Live Sanely and Simply in a Troubled World.* Schocken Books, 1970.

Odum, Howard T., and Elisabeth C. Odum. *Energy Basis for Man and Nature.* McGraw-Hill Companies, 1976.

Odum, Eugene P. *Fundamentals of Ecology.* W. B. Saunders Company, 1971.

Schwenk, Theodore. *Sensitive Chaos: The Creation of Flowing Forms in Water and Air.* Rudolf Steiner Press, 1976.

Smith, J. Russell. *Tree Crops: A Permanent Agriculture.* Harcourt, Brace and Co., 1929.

Van der Ryn, Sim, and Stuart Cowan. *Ecological Design.* Island Press, 1996.

Wackernagel, Mathis, and William Rees. *Our Ecological Footprint: Reducing Human Impact on the Earth.* New Society Publishers, 1998.

Farming, Food, and Water

Ashworth, Suzanne. *Seed to Seed: Seed Saving and Growing Techniques for Vegetable Gardeners.* Seed Savers Exchange, 2002.

Brunetti, Jerry. *The Farm as Ecosystem.* Acres U.S.A., 2014.

Buchmann, Steven L., and Gary Paul Nabhan. *The Forgotten Pollinators.* Island Press, 1996.

Coleman, Eliot. *Four-Season Harvest: Organic Vegetables from Your Garden All Year Long.* Chelsea Green Publishing, 1999.

Crawford, Martin. *Creating a Forest Garden: Working with Nature to Grow Edible Crops.* UIT Cambridge Ltd., 2010.

Deppe, Carol. *The Resilient Gardener: Food Production and Self-Reliance in Uncertain Times.* Chelsea Green Publishing, 2010.

Douglas, James Sholto, and Robert Adrian de Jauralde Hart. *Forest Farming: Towards a Solution to Problems of World Hunger and Conservation.* Rodale Books, 1976.

Falk, Ben. *The Resilient Farm and Homestead: An Innovative Permaculture and Whole Systems Design Approach.* Chelsea Green Publishing, 2013.

Fern, Ken. *Plants for a Future: Edible and Useful Plants for a Healthier World.* Permanent Publications, 1997.

Flores, H. C. *Food Not Lawns: How to Turn Your Yard into a Garden and Your Neighborhood into a Community.* Chelsea Green Publishing, 2006.

Fukuoka, Masanobu. *Natural Way of Farming: The Theory and Practice of Green Philosophy.* Japan Publications, 1985.

——. *The One-Straw Revolution: An Introduction to Natural Farming.* Rodale Books, 1978.

Hart, Robert. *Forest Gardening: Cultivating an Edible Landscape.* Chelsea Green Publishing, 1996.

Jeavons, John. *How to Grow More Vegetables Than You Ever Thought Possible on Less Land Than You Can Imagine,* 5th ed. Ten Speed Press, 1995.

Jenkins, Joseph C. *The Humanure Handbook: A Guide to Composting Human Manure,* 3rd ed. Jenkins Publishing, 2005.

Kourik, Robert. *Designing and Maintaining Your Edible Landscape Naturally.* Chelsea Green Publishing, 2005.

——. *Roots Demystified: Change Your Gardening Habits to Help Roots Thrive.* Metamorphic Press, 2007.

Logsdon, Gene. *Getting Food from Water: A Guide to Backyard Aquaculture.* Rodale Press, 1978.

Ludwig, Art. *Create an Oasis with Greywater: Your Complete Guide to Choosing, Building and Using Greywater Systems.* Oasis Design, 2000.

——. *Water Storage: Tanks, Cisterns, Aquifers, and Ponds for Domestic Supply, Fire and Emergency Use– Includes How to Make Ferrocement Water Tanks.* Oasis Design, 2005.

Mudge, Ken, and Steve Gabriel. *Farming the Woods: An Integrated Permaculture Approach to Growing Food and Medicinals in Temperate Forests*. Chelsea Green Publishing, 2014.

Phillips, Michael. *The Holistic Orchard: Tree Fruits and Berries the Biological Way*. Chelsea Green Publishing, 2012.

Shepard, Mark. *Restoration Agriculture: Real-World Permaculture for Farmers*. Acres U.S.A., 2013.

Solomon, Steve. *The Intelligent Gardener: Growing Nutrient-Dense Food*. New Society Publishers, 2013.

Stamets, Paul. *Mycelium Running: How Mushrooms Can Help Save the World*. Ten Speed Press, 2005.

Toensmeier, Eric. *The Carbon Farming Solution*. Chelsea Green Publishing, 2016.

Whitefield, Patrick. *How to Make a Forest Garden*. Permanent Publications, 2002.

Yeomans, P. A., and Ken. *Water for Every Farm*. Keyline Designs, 1993.

Home and Energy

Kern, Ken. *The Owner-Built Home*. Owner-Builder Publications, 1972.

Olkowski, Helga, Bill Olkowski, and Sim Van der Ryn. *The Integral Urban House: Self-Reliant Living in the City*. Sierra Club Books, 1979.

Stoyke, Godo. *The Carbon Busters Home Energy Handbook: Slowing Climate Change and Saving Money*. New Society Publishers, 2007.

Economy and Finance

Greer, John Michael. *The Wealth of Nature: Economics as if Survival Mattered*. New Society Publishers, 2011.

McKibben, Bill. *Deep Economy: The Wealth of Communities and the Durable Future*. Henry Holt and Company, LLC, 2007.

Roland, Ethan, and Gregory Landua. *Regenerative Enterprise: Optimizing for Multi-Capital Abundance. Version 1.0*. Lulu.com, 2013.

Community and Landscape

Anderson, Kat. *Tending the Wild: Native American Knowledge and the Management of California's Natural Resources*. University of California Press, 2006.

Corbett, Judy, and Michael Corbett. *Designing Sustainable Communities: Learning from Village Homes*. Island Press, 2000.

Cronon, William. *Changes in the Land: Indians, Colonists, and the Ecology of New England*. Hill and Wang, 2003.

Donahue, Brian. *Reclaiming the Commons: Community Farms and Forests in a New England Town*. Yale University Press, 1999.

Todd, Nancy Jack, and John Todd. *Bioshelters, Ocean Arks, City Farming: Ecology as the Basis of Design*. Sierra Club Books, 1984.

Weisman, Alan. *Gaviotas: A Village to Reinvent the World*. Chelsea Green Publishing, 1998.

Regenerative Design and Resilience

Lyle, John Tillman. *Design for Human Ecosystems: Landscape, Land Use, and Natural Sources*. Van Nostrand Reinhold, 1985.

——. *Regenerative Design for Sustainable Development*. John Wiley & Sons, Inc., 1994.

McDonough, William, and Michael Braungart. *Cradle to Cradle: Remaking the Way We Make Things*. North Point Press, 2002.

Todd, Nancy Jack, and John Todd. *From Eco-Cities to Living Machines: Principles of Ecological Design*. North Atlantic Books, 1994.

Walker, Brian, and David Salt. *Resilience Thinking: Sustainable Ecosystems and People in a Changing World*. Island Press, 2006.

Zelov, Chris, and Phil Cousineau. *Design Outlaws on the Ecological Frontier*. Knossus Project, 2000.

New Visions of Ecology and Invasive Species

Marris, Emma. *The Rambunctious Garden: Saving Nature in a Post-Wild World*. Bloomsbury USA, 2013.

Orion, Tao. *Beyond the War on Invasive Species: A Permaculture Approach to Ecosystem Restoration*. Chelsea Green Publishing, 2015.

Pearce, Fred. *The New Wild: Why Invasive Species Will Be Nature's Salvation*. Beacon Press, 2015.

Thompson, Ken. *Where Do Camels Belong?: The Story and Science Behind Invasive Species*. Profile Books, 2014.

Permaculture Guidebooks

Hemenway, Toby. *Gaia's Garden: A Guide to Home-Scale Permaculture*. Chelsea Green Publishing, 2000.

Morrow, Rosemary. *Earth User's Guide to Permaculture*. Kangaroo Press, 2007.

Magazines

Permaculture Design magazine (formerly *Permaculture Activist*)
permaculturedesignmagazine.com
Permaculture magazine (UK)
permaculture.co.uk
Permaculture, North America
permaculturemag.org

THE SHOULDERS WE STAND ON

Sir Isaac Newton is believed to have said "If I have seen further, it is by standing on the shoulders of giants." Even such an inventor, philosopher, and maybe the most influential scientist in history – who described the laws of gravity – recognized his dependence on the people who came before him, the people whose findings and understanding he built from and utilized in his own research. In the same way, the work happening now in ecological design, permaculture, homesteading, sustainable living skills, agroforestry, and ecology is building on the important work and the trail that was broken by numerous people within numerous cultural traditions.

In the 1994 film *Ecological Design: Inventing the Future*, where many of the foremost ecological designers of the recent past pay tribute to Buckminster Fuller, there is a point at which Fuller goes into the Pillow Dome Bioshelter erected at the New Alchemy Institute formerly in Falmouth, Massachusetts. Made of metal and using the principles of the geodesic dome "Bucky" invented, he lies down on the bed inside the dome and stares up in peace and wonder. He says "This is my dream made real." It is a poignant moment where the elder has passed his knowledge to the next generation and they have taken his inspiration and knowledge and built upon it to create new knowledge and understanding. In this case they have taken ideas and vision from Bucky and built it in the world, making it real and something we can learn from even more.

The work we see today, the pioneers and the inventors, the groups developing permaculture sites, courses, innovative practices and bringing these ideas into many communities are the latest to take up the mantle and carry forward the important ideas and information about permaculture. All of their work (and my own) builds on previous work.

Permaculture owes a debt of gratitude for its development not only to Bill Mollison and David Holmgren, who tied together the bundle of many ideas and made the connections and interrelationships that make it such a powerful perspective and design approach, but also to myriad people in the 1960s and 1970s who were working on revolutionary responses to the ills of our times. Included among them are: the Farralones Institute, founded in the 1970s by Sim Van der Ryn; Donella "Dana" Meadows, lead author of *The Limits of Growth* (1971) and founder of the Sustainability Institute; Amory and Hunter Lovins of the Rocky Mountain Institute; John Tillman Lyle's Institute for Regenerative Design; and John and Nancy Jack Todd, cofounders of the New Alchemy Institute. And there are the people who inspired that generation of innovators and teachers: J. Russell Smith and his work on tree crops; Helen and Scott Nearing, who inspired a generation of homesteaders; and J. I. Rodale, early advocate for sustainable farming and organic agriculture and founder of the Rodale Institute.

We are all truly indebted to and should be thankful for the shoulders we stand on.

GRATITUDE

I want to thank the network of permaculture practitioners working everywhere to change the world for the better. Thanks to my first permaculture teachers, Jude Hobbs, Tom Ward (a.k.a. Tomi Hazel), Rick Valley, and the Pacific Northwest network. Thanks to my teachers and colleagues around the Northeast region: Dave Jacke, Eric Toensmeier, Jonathan Bates, Lisa DePiano, Claudia Joseph, Lisa Fernandes, and many others.

Thanks to Carleen Madigan for her insight, inspiration, and help in shepherding this book through the editing process. Thanks to Bas Gutwein, Keith Zaltzberg, and the rest of the Regenerative Design Group team. Thanks to Sue Bridge for her inspiration and spirit, and for making Wildside Cottage and Gardens available as a learning laboratory.

Thanks to Kemper, my steady partner of many years. And thanks to my son, Forest, and all of his generation for carrying these ideas into the future, whatever it brings.

INDEX

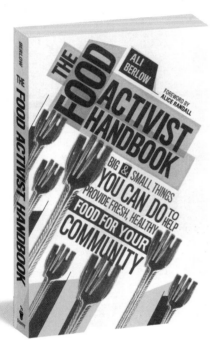
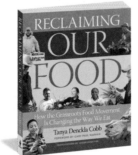